EXPERIMENTAL METHODS
AN INTRODUCTION TO THE ANALYSIS AND PRESENTATION OF DATA

EXPERIMENTAL METHODS

AN INTRODUCTION TO THE ANALYSIS AND PRESENTATION OF DATA

LES KIRKUP

JOHN WILEY & SONS, INC.

BRISBANE • NEW YORK • CHICHESTER • WEINHEIM • SINGAPORE • TORONTO

First published 1994 by
JACARANDA WILEY LTD
33 Park Road, Milton, Qld 4064

Offices also in Sydney and Melbourne

Typeset in 10.5/12 pt Garamond

National Library of Australia
Cataloguing-in-Publication data

Kirkup, L. (Les).
   Experimental methods: an introduction to the
   analysis and presentation of data.

   Include index.
   ISBN 0 471 33579 7.

   1. Physics — Laboratory manuals. 2. Physics
   — Experiments. 3. Physics — Data processing.
   I. Title.

530.078

Cover photograph: APL/Orion Press

All the experiments described in this book have been written
with the safety of both teacher and student in mind. However,
all care should be taken and appropriate protective clothing
should be worn when carrying out any experiment. Neither the
publisher nor the author can accept responsibility for any
experiments described in this book.

Printed in Singapore

10 9 8 7 6 5 4

To Janet and me mam

CONTENTS

PREFACE

As an organiser and demonstrator in undergraduate laboratories, I have never been entirely satisfied with the 'ad hoc' arrangements made to assist those new to experimentation to enable them to come to grips with vital topics such as uncertainties, graphing and keeping a laboratory notebook. It was this dissatisfaction coupled with a belief that other topics such as report writing and the use of computers in the gathering, presentation and analysis of experimental data should be given more emphasis in texts concerned with experimental methods, that encouraged me to put pen to paper.

Experimental work should be amongst the most stimulating and satisfying activities in any course of study in science or engineering. A well-designed and executed experiment followed by *proper data analysis* offers insights into a process or phenomenon which are unlikely to be provided by a theoretical study alone. It is what goes into the *proper analysis* of experimental data that dominates the subject matter of this book. I have tried to describe the tools and techniques that will assist those engaged in presenting and analysing experimental data as part of their undergraduate studies in science or engineering.

This book is aimed firstly at those beginning an undergraduate course of study in the physical sciences or engineering who have a significant laboratory component to their subjects. Early chapters assume very little prior knowledge of the techniques of data analysis. However, several topics in the book go beyond first-year level and many of the matters raised should be of relevance to those engaged in experimental work up to the senior stages of an undergraduate course.

With regard to the formulae that appear in the text, I have preferred to emphasise the background and assumptions relevant to a particular formula rather than to deal in-depth with the mathematics of its derivation. Wherever possible, discussion of a formula is followed up with an appropriate 'worked' example. I have given references in the text to where greater details concerning formula derivation can be found.

The book is divided into three sections:

(i) Chapters 1 to 4 cover the basic groundwork appropriate to first-year studies in science and engineering in the areas of notebook keeping, characterising experimental data, uncertainties and graphing.

(ii) Chapters 5 to 7 discuss the statistical analysis of experimental data and the important topic of least-squares fitting of functions to data as well as addressing the crucial area of communicating the findings of an experiment in the form of a report. Though 'long' reports are generally given more emphasis beyond first year, I feel that it is a topic of such importance that it deserves fuller treatment than a brief mention of the headings found in a report. The material contained in these chapters is likely to be most applicable to data-analysis problems beyond first-year undergraduate level.

(iii) Chapters 8 and 9 deal with the application of technology to the analysis and presentation of data. Though pocket calculators have been widely available for more than twenty years, their usefulness for analysing experimental data is rarely dealt with explicitly in a book of this type. Superseding the calculator in power is the microcomputer, and the availability of excellent hardware and software for presentation and analysis purposes has encouraged me to write about these matters.

Because of the diverse range of topics that appear in this book I have included, in an appendix, details of other books which can be referred to for more information. As microcomputers are now commonly found in undergraduate laboratories assisting with data gathering as well as analysis, the last appendix offers an introduction to this topic.

I have gained enormous enjoyment from writing this book (and learned a few things along the way!) and look forward to any feedback it may provoke. Suggestions from students have been mostly heeded (more examples, more examples!) and I hope this has made for a book more in keeping with their needs.

There are many people who have helped make this book possible. In particular I would like to thank my publishing editor, Derelie Evely, for her enthusiastic support of the project and her deadlines. Equal thanks are owed to my university for granting me the time away from my normal duties to complete the book, and to Professor David Rawson of the University of Luton, England for providing a congenial environment in which to carry out some of the work. I'd also like to thank the following reviewers and providers of ideas for examples, experimental data and problems.

Brian McInnes (University of Sydney),

Graham Russell (University of New South Wales),

Roger Rassool (University of Melbourne),

George Haig, Leon Firth (University of Paisley)

Barry Haggett, Bill Roe, John Plater, Rosalyn Butler (University of Luton),

Bob Cheary, Jeff Kershaw, John Bell, Maree Gosper, Nick Armstrong, Patsy Gallagher, Peter Logan, Ray Woolcott, Walter Kalceff (University of Technology, Sydney).

In addition I would like to express my gratitude to Janet Sutherland, Shona Rawson and Billy Ward for many helpful suggestions.

Les Kirkup
University of Technology, Sydney
e-mail: kirkup@phys.uts.edu.au

1 INTRODUCTION TO EXPERIMENTATION

1.1 Overview: the importance of experiments in science and engineering

Scientists and engineers devote a large amount of time to what might be termed 'experimental work'. What are the reasons for doing experiments? In the first place, scientific and technical advances rely heavily on the support that a critical experiment, or series of experiments, can give. New and old theories are 'put to the test' through experiments. Devising and executing an experiment that provides a thorough test of a theory may not be easy, but until such a test is undertaken, and the results confirmed independently by others, the theory is unlikely to gain complete acceptance. Additionally, carefully performed experiments may reveal new effects that require existing explanations to be modified.

On another level, experiments performed as part of a laboratory course are unlikely to break new ground. However, they provide us with the opportunity to acquire knowledge, skills and understanding through investigating the 'real world'. This has advantages over the idealised descriptions and explanations of phenomena often presented in textbooks: seeing something happen has much more impact than reading about it. *Making* something happen, as we do when we perform an experiment, leaves an even greater impression on the mind.

Experiments are not without their difficulties; some experimental techniques take time to master and occasionally we are confronted with a mass of data that requires careful examination before we are able to draw out the important features. In these circumstances a little patience and perseverance go a long way.

While performing an experiment, you may need to acquire proficiency in the operation of equipment of varying degrees of sophistication. It is not the purpose of this book to give instruction on the

1

operation of particular instruments, but to offer advice of a more general and, we hope, enduring nature. Specific examples and exercises are included within the forthcoming chapters. These should assist in illustrating a particular analysis technique or reinforcing an important point.

An experiment may be required to assist in clarifying a number of questions, some of a general nature and others more specific. As examples:

 (i) How widely applicable is a particular theory?
 (ii) At what temperature does a newly prepared alloy melt?
 (iii) Can a technique for measuring a particular quantity be improved upon?
 (iv) What happens to the magnetic properties of a material when it is cooled to very low temperatures?
 (v) Does the bombardment of a semiconductor device with nuclear radiation affect its performance?

It is these types of questions that form the starting point for experimental investigations.

It is instructive to outline the stages through which a typical experiment develops. Though the stages are presented in a particular order, there can be much moving back and forth between stages, as ideas change, theories are modified, better apparatus becomes available, and so on.

1.2 Stages of a typical experiment

The aim

This is the starting point of an experiment. What do we want to find out? The clearer and more well defined the aim of the experiment, the easier it is to do the planning to achieve that aim. The aim may contain an idea or hypothesis that you want to advance or test. However, it is not unusual to begin with a particular objective in mind and while doing the experiment discover something interesting and unexpected. This is the nature of experimentation. If the outcomes of an experiment were wholly predictable there would be little point in undertaking them. However, we must be aware of the risk of becoming side-tracked and failing to complete the original work.

The plan

Once the aim has been decided, a plan is devised for achieving the aim. Decisions are made as to what equipment is required, which quantities need to be measured, and the manner by which they are to be measured.

Preparation

The preparation stage involves organising the experiment. Equipment is collected and assembled. If the experimental technique or instrument to be used is unfamiliar, instruction should be taken from an experienced user. This avoids wasting time and making obvious blunders, as well as reducing the chance of damaging the equipment (and possibly yourself!).

Preliminary experiment

Once the equipment has been assembled, a preliminary experiment is often performed. This promotes familiarity with the operation of the equipment, indicates which features work well and which need further development, and gives a feel for what values to expect when the experiment is performed more carefully.

Collecting data

The data collection phase now begins. Alertness and attention to detail at this stage will reward you with a valuable set of data. There is nothing more frustrating than spending an afternoon collecting data, only to find that something, such as neglecting to record the units in which the measurements were made, has rendered the data unusable.

Repeatability

The experiment is carefully repeated in order to verify whether the first set of data is representative and can be reproduced one or more times. The repeated experiment cannot be expected to generate *exactly* the same data. However, gross variations between sets of data need to be investigated.

Analysis of data

When data collection is complete, a most important question is asked: 'What do the data tell me?' If the experiment was performed with some hypothesis in mind, then this will indicate what data analysis method(s) should be adopted. For example, in an experiment to study the change in light intensity as the distance from the light source increases, a prevailing theory may lead us to believe that there is a 'power law' relationship between intensity, I, and distance, d. The power law relationship could be written:

$$I = Ad^n, \tag{1.1}$$

where A and n are constants. The experiment would consist of measuring the light intensity as the distance between the light source and detector is varied. From the analysis of the data we would like to know how well equation 1.1 describes the relationship between I and d, and the numerical values for A and n.

What do the data tell you?

Once the data have been gathered and analysed, it is time to decide whether they are consistent with the initial hypothesis or whether the evidence is inconclusive or even contradictory. For example, in our light experiment, do the data provide us with enough evidence to be able to conclude that equation 1.1 *is* a good description of the relationship between intensity and distance?

Reporting the experiment

When the work in the laboratory is completed, the findings should be communicated in a clear and concise way. A report may be prepared which describes the important features of the experiment such as the aim, method, data, analysis and conclusion[1].

1.3 Keeping a record of your work

An experiment may take as little time as an hour or extend over days. During this time, procedures may be devised, electrical circuits built, apparatus assembled, data gathered and other steps taken, large and small, before the experiment is complete. Irrespective of the duration or complexity of the experiment, one thing is certain: the better the record that has been made of what has been done, the easier will be the task of presenting the work, perhaps in the form of a report to a laboratory supervisor. A convenient way of recording work is to use a laboratory notebook, sometimes referred to as a 'log book'.

1.3.1 The laboratory notebook

A laboratory notebook contains a permanent record of experiments performed, and for many scientists and engineers is an indispensable element of experimental work. In it they record every detail of an experiment, whether or not those details seem important at the time. Often it is not until sometime later that it emerges which are the important entries, and which are of less value.

Notebooks are available in various sizes, but those that have alternate pages of lined paper followed by graph paper are convenient

---

1. Report writing is discussed in chapter 7.

to use in situations in which a graph needs to be plotted as part of the experiment. An alternative is to use a standard lined notebook and fix in graph paper when necessary. A hardback notebook, though generally more expensive than the softback variety, is a good investment, especially as it tends to come in for some harsh treatment and may be required to last for a long time. In addition, there is generally less tendency for people to tear out pages from a hardback book (an act to be frowned upon under all circumstances!).

A laboratory notebook needs to be intelligible to at least one person — you! However, there may be situations in which the contents of the notebook form part of an assessment of an experiment. In this case you need to remember that the book is going to be read by someone else, so a logical layout of the account of the experiment is recommended. The order and description of important elements that make up an account of an experiment are given in table 1.1.

Table 1.1: Description of notebook contents

Notebook entries	Description
Date	This is good housekeeping and allows you to chronologically tie the contents of your notebook to other parts of your work. For example, it might be important to associate entries in your book with experimental data that you gathered and stored on computer on the same day.
Title	Leave the reader in no doubt as to what the experiment is about!
Aim of the experiment	Though this is something that may have already been decided on your behalf, it is *so* important that it bears repetition and should be given a prominent place in your notebook after the title of the experiment.
Description of the apparatus	For many experiments a brief list of the apparatus used is sufficient. If this has already been included on an 'experimental method' sheet supplied at the beginning of the experiment, then a sensible option is to attach that sheet in a permanent manner to your notebook. Recording the serial numbers of instruments you use is good practice. This permits you to return to an instrument later if analysis reveals irregularities which you have reason to suspect were caused by that instrument.

(continued)

Table 1.1 (*continued*)

Notebook entries	Description
Sketch of apparatus	A fully labelled diagram of the experimental arrangement is worth a page of explanation and can assist in recalling the experiment long after it has been completed. A simple line diagram drawn freehand is all that is required.
Experimental method	If the method is given in the form of a set of instructions to follow, then they can be 'cut and pasted' into the notebook. If you devise the method yourself, then give enough detail so that you, or someone else, could repeat the experiment at a later date.
Measurements	It is usual to present data in tables, taking care to give each column in the table a heading which includes the unit of the quantity measured. An estimate of the experimental uncertainty (also known as experimental error) in each quantity should be included in the table. Measured values are recorded *directly* into the notebook as they are made.
Graphs	Graphs are better for giving you 'the big picture' of the data than a table of numbers, and are often the first thing that someone assessing your work will look at. The graph must be presented properly with title, labelled axes, etc.
Calculations	If you need to make calculations based on your data, clearly state the formula or relationship you are going to use. Work through the calculation as fully as possible in your notebook, giving a clear account of the steps you have taken.
Conclusion	A laboratory notebook is not usually the place where you will present a detailed discussion of your work. However, at the end of the experiment you should include a brief conclusion. For example, if the aim of the experiment was to find the melting point of lead, you might report: *From the observations made in this experiment, the melting point of lead is found to be $(325 \pm 5)°C$.*

A useful tip is to number the pages of the notebook. This is helpful when you want to refer to another piece of work in the book and is especially profitable when describing a series of experiments covering many pages. For example, you might want to say '*the circuit used is that shown on page 27*'.

As described above, the notebook is really performing two functions. The first is to record all the relevant information concerning the experiment, and the second is to present it as a mini-report that might be assessed by others. Although both functions are very important, as the length and complexity of experiments increase, these functions tend to become separated and the role of the notebook changes considerably.

1.3.2 Advanced experiments and the laboratory notebook

So long as experiments are short and reasonably self-contained, the approach to presenting the work in your notebook described in the above section is fine. But what happens if you are asked to devise an experiment from scratch, which demands you be responsible for the aim and method as well as gathering data, assessing uncertainties and so on? This situation may well occur when doing more advanced experiments or when performing an experimental project lasting a few weeks. Though the section headings appearing in the left column of table 1.1 are still helpful, it is unlikely that you will be able to fill out the notebook in such a stepwise manner.

A typical situation you might find yourself in could be as follows.

As part of an experiment you need to measure the size of the magnetic field in the vicinity of your apparatus. You decide to use a Hall probe[2] and you set about calibrating it. When you try to use it for your experiment, it is found to be too insensitive for measuring the small fields you are confronted with and another approach is required. On looking through books on electromagnetism you come across a description of a search coil, which generates a voltage when it is rotated in a magnetic field. Using this and an electronic integrating circuit you believe you will be able to measure very small magnetic fields. You decide to wind a search coil and test out the circuit.

The work described above may take a week of testing equipment, building circuits, seeking information and going down the odd 'dead end'. Throughout this development stage everything is recorded in the laboratory notebook: results of the calibration of

---

2. A Hall probe consists of a small piece of semiconductor which generates a small voltage proportional to the magnetic field in the vicinity of the probe.

the system, unsuccessful attempts at measuring the magnetic field in the vicinity of the apparatus, description of the design of the search coil and so on. When ideas occur about some other way of proceeding, they are recorded in the notebook. It may be that only a few of the ideas will ultimately contribute to the current experiment. However, it is quite possible that, much later, you will be confronted with a similar problem and your recorded ideas, references and measurements will provide a good starting point for solving *that* problem.

We have described in the indented text on page 7 a small section of a long experiment. The full experiment may take some time to complete, and the account of it could fill a laboratory notebook. In this case, the notebook serves a crucial role when the work that has been done is to be reported to others. It contains details of the progress made, problems that were encountered, as well as data gathered in tabular and graphical form. Writing a report is so much easier when all the details you require are in one place — your laboratory notebook. It is unlikely that the notebook by itself will make complete sense to anyone else, but it must make sense to *you* as it is from the notebook that a report which *will* be read by others can be prepared.

If you are not familiar with laboratory notebooks already, you might be curious as to what the contents of a typical notebook look like. Figure 1.1 illustrates a short section from one of my notebooks dating back to 1992. In fact, the work relates to magnetic field measurement mentioned at the beginning of this section. You will appreciate that neatness was not the top priority, but the method by which a sensitive measurement of magnetic field may be made can be seen taking shape. Contained in the section are details of the coil, the basic electromagnetic theory which leads to a formula relating the integrated voltage with time from an electronic circuit to the change in magnetic field. A 'textbook' integrator is shown along with my first attempt at a practical version. Where thoughts have occurred to me concerning the design of the circuit, I have recorded them in the notebook. The table shows the drift of output voltage of the integrator with time over a period of 60 s.

A laboratory notebook is a very personal item and in some circumstances can be unexpectedly valuable. If you are lucky enough to make a major scientific or technical discovery, it is likely to appear in your notebook before anywhere else. If, in addition, you are *unlucky* enough to be in competition with someone else who claims to have made the discovery first, you might find that your notebook becomes a vital piece of legal evidence to show what you did and when. Beware: money and reputation could be tied to your laboratory notebook, so use it well and keep it safe!

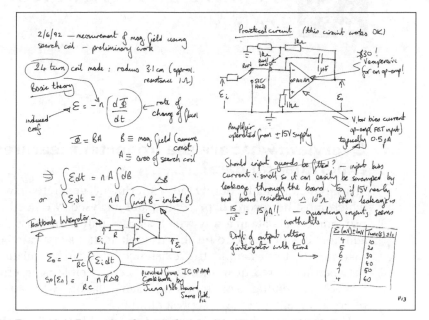

Figure 1.1: Example of pages from a laboratory notebook

2 CHARACTERISTICS OF EXPERIMENTAL DATA

2.1 Overview: what are the important features of experimental data?

Measurements made during an experiment generate 'raw' data which must be recorded, presented and analysed. We need to display numerical data in a way which assists analysis of those data. In addition, it is desirable to be able to look at the data as a whole so that trends, for example the existence of a linear relationship between measured quantities, can be seen. A table is an effective way of presenting data requiring manipulation, while a graph offers a revealing pictorial representation.

There are a number of questions to consider while gathering experimental data:

(i) What is the unit associated with each measurement?
(ii) How much variability is there in the data?
(iii) Can we estimate the size of a quantity before it is measured?
(iv) What should be included in a table of data?

We will now turn our attention to these and related questions.

2.2 Units of measurement

At the heart of an experiment lies measurement and measurement requires a system of units. When a scientist or engineer claims to have made a breakthrough, perhaps discovering a new ceramic material with remarkable electrical properties at low temperatures, other workers in that area of science or engineering want to know as many details of the characteristics of this material as possible. These might include its melting point, density, electrical conductivity, heat capacity and crystal structure.

Above all, *quantitative* estimates of the properties are needed. These permit measurements made in other laboratories (perhaps in another part of the world) to be compared directly with the original work. A starting point for that comparison is that everyone agrees on a set of units to measure quantities in, and uses those units consistently when the results of experiments are reported.

2.2.1 The SI system of units

Perhaps the most widely used system of units is the Systeme International, or SI system, and it is the one we will use throughout this book.[1] At its core there are seven fundamental units, which are shown in table 2.1.

Table 2.1: Fundamental SI units

Quantity	Name of unit	Symbol
mass	kilogram	kg
length	metre	m
time	second	s
electrical current	ampere	A
temperature	kelvin	K
luminous intensity	candela	cd
amount of substance	mole	mol

Other units are derived from the fundamental units. An example of a derived unit is the metre per second (written $m\ s^{-1}$ or m/s) which is the unit of velocity.

For a detailed description of the fundamental units, how they are defined, and how they may be combined to give a host of other units, a good book to refer to is *Handbook of Units and Quantities* (Rocke, 1984) (full details in appendix 1). Many of the derived units have been given names which may be familiar to you. As examples, the newton, volt and joule are all derived units.

In pursuit of our goal of effective presentation and analysis of data, we should remember: *Whenever we fill a table with data, plot a graph or make a remark concerning measurements or calculations based on those measurements, we must **always** state the units in which we are working.*

This can be put another way. Tables, graphs, calculations and conclusions are *meaningless* unless the units of the quantities are

---

1. We will not use SI units exclusively. Units such as the minute and the degree are so commonly used that examples will be given using these units.

stated. A critical question we need to ask when about to use an instrument is 'In what unit does this instrument measure something?' When using a stopwatch the answer may be obvious. When using less familiar apparatus, such as a vacuum gauge, the answer may be less apparent.

2.2.2 Multiples and subdivisions of units

For some measurements, the fundamental unit might be clumsy and it would be more convenient to use a multiple or subdivision of that unit. An example can be taken from the topic of electricity. The farad (symbol F) is the SI unit of electrical capacitance. However, it is unlikely that you will come across a capacitor with a capacitance as large as 1 F. Much more commonplace are capacitors with values around 10^{-6} F or less. It would be convenient if we could assign a name to the subdivision 10^{-6} F, and in fact there already is a name: it is the *micro*farad (symbol μF). It is so much easier to speak of 'five point five microfarads' than to have to say 'five point five times ten to the minus six farads'! As 'micro' is placed in front of the unit, it is termed a *prefix*.

There are prefixes for multiples of units extending from 10^{-24} to 10^{24}. Table 2.2 shows the prefixes corresponding to each multiplier in the range 10^{-15} to 10^{12}, which will cover most situations you are likely to encounter.

Table 2.2: Si prefixes

Power of ten	Prefix	Symbol	Example
10^{-15}	femto	f	fs (femtosecond $\equiv 10^{-15}$ s)
10^{-12}	pico	p	pF (picofarad $\equiv 10^{-12}$ F)
10^{-9}	nano	n	nA (nanoampere $\equiv 10^{-9}$ A)
10^{-6}	micro	μ	μPa (micropascal $\equiv 10^{-6}$ Pa)
10^{-3}	milli	m	mJ (millijoule $\equiv 10^{-3}$ J)
10^{-2}	centi	c	cm (centimetre $\equiv 10^{-2}$ m)
10^{3}	kilo	k	kV (kilovolt $\equiv 10^{3}$ V)
10^{6}	mega	M	MW (megawatt $\equiv 10^{6}$ W)
10^{9}	giga	G	GHz (gigahertz $\equiv 10^{9}$ Hz)
10^{12}	tera	T	TΩ (teraohm $\equiv 10^{12}$ Ω)

Note that, with the exception of centi, all the other powers of ten that appear in table 2.2 are multiples of three.

Incidentally, while we find it convenient in many situations to express length in centimetres, it is unusual to see the centi prefix used elsewhere. For example, it is rare to hear a voltage expressed as '2.2 centivolts', or a change in temperature as '50 centikelvins'.

EXERCISE A

Express:

(i) 2.2×10^{-6} V in μV

(ii) 6.2×10^{-2} m in mm

(iii) 6.52×10^{4} J in kJ

(iv) 1.8×10^{5} W in MW

(v) 6.7×10^{-11} F in pF

2.3 Tabulation of data

Experimental observations take forms as contrasting as finding the distance to the moon using laser interferometry, to measuring the temperature of an oil bath using a mercury-in-glass thermometer. Whatever the measurement technique or instrument used, carefully made observations are the cornerstone of good experimental work. It is important that they are properly recorded in a manner that will permit analysis later.

There are two types of experimental situation that commonly occur. In the first, repeated measurements of a quantity are made where there is no reason to believe that the quantity is varying. Examples of this would be timing how long it takes an object to fall through a given distance or measuring a particular wavelength of light emitted from a lamp containing helium gas. In the other situation, we are trying to establish the relationship between two quantities — call them A and B. This we would do by changing A and observing the effect that this has had on B. An example of this would be measuring the viscosity of a liquid (quantity B) as the temperature of the liquid changes (quantity A).

In both situations a convenient way in which to present data, whether it be in a laboratory notebook or a report of the experiment, is in the form of a table. Table 2.3 contains the values of ten repeated measurements of the time it took a small object to fall through a distance of 25 m.

Table 2.3: Times for an object to fall 25 m

Time of fall (s)	2.2	2.0	2.6	1.9	2.1	2.4	2.2	2.3	2.3	2.0

For the table to be useful, it needs a distinct heading with a clearly stated unit. The unit is shown in brackets after the name of the quantity being measured. In many circumstances, it is also useful to estimate the uncertainty in each measurement as it is made (as discussed later in section 2.4) and to include that information in the table.

An alternative method of indicating the unit of measurement in a table is shown in table 2.4.

Table 2.4: Times for an object to fall 25 m

Time of fall/s	2.2	2.0	2.6	1.9	2.1	2.4	2.2	2.3	2.3	2.0

The argument for using /s instead of (s) is as follows: tables can contain only numbers. In this experiment we have measured time intervals in seconds, so in order to be able to display those intervals as numbers in the table we must 'cancel out' the unit by *dividing* our measurements by seconds, hence the /s next to the name of the quantity. This approach to indicating units, although favoured by many authors, can sometimes be confusing and I prefer the method of presenting units within brackets in the heading of the table.

As an experiment proceeds, the measured values are recorded in a table with no attempt at data manipulation, such as squaring values or subtracting a constant. If you *do* modify the data as you go along, and in the process make a mistake, it can be very difficult to work back to find out what went wrong. A classic example of this is when converting from one form of unit to another 'in your head' and wondering later why the final answer is a thousand times too large (or too small).

2.3.1 Tabulating data represented in scientific notation

Writing the length 1.1 μm as 1.1×10^{-6} m is to express the value in *scientific notation*[2] (see section 2.5.3 for more details). If we have data expressed in scientific notation which we wish to enter into a table, we improve the readability of the table by indicating the multiplying power of ten in the heading of the appropriate row or column of the table.

For example, as part of an experiment the electrical inductance of a coil of wire was measured. The following data were obtained:

9.5×10^{-3} H, 9.3×10^{-3} H, 9.9×10^{-3} H, 9.9×10^{-3} H, 9.1×10^{-3} H

---

2. Sometimes this is referred to a 'powers of ten' notation.

Table 2.5 displays the data by showing the multiplying factor, expressed as a power of ten in the heading of the table.

Table 2.5: Measured values of the inductance of a coil of wire

Inductance ($\times 10^{-3}$ H)	9.5	9.3	9.9	9.9	9.1

EXERCISE B

Present the following pressure data in a table in the manner described in this section.

1.03×10^5 Pa, 1.01×10^5 Pa, 1.01×10^5 Pa, 9.9×10^4 Pa, 1.05×10^5 Pa, 1.08×10^5 Pa

2.4 Uncertainties in measurements

Despite our best efforts or the quality of the equipment we use, there is going to be an amount of variability in quantities measured in an experiment. It could be that during a measurement of temperature of a liquid using a mercury-in-glass thermometer, the position of the top of the mercury column fluctuates slightly. We may have difficulty deciding whether the mercury is at the 15.5°C mark or has reached the 16.0°C mark. We shouldn't feel frustrated that we cannot make an 'exact' measurement. Such things don't exist. We might need to look at the design of the experiment to try to ensure that we have done all we can to measure the important quantities as accurately as possible.

Nevertheless, some *uncertainty* in the measured quantity must remain no matter how good the experimenter or experimental design. (Experienced experimenters are constantly trying to find ways to reduce the uncertainty in the measurements they make.)

Suppose the fluctuating position of the top of the mercury column in the thermometer means that no individual temperature measurement can be made to better than 0.5°C. We can write that the temperature of the liquid is:

$$(15.5 \pm 0.5)°C$$

The ± sign indicates we believe that the temperature lies in the interval $(15.5 - 0.5)°C$ to $(15.5 + 0.5)°C$, that is, somewhere between 15.0°C and 16.0°C.

Estimates of the uncertainty in measurements should always accompany the measurement and need to be recorded in your laboratory notebook. Often the best place to write the uncertainty, especially if it is the same for all measurements, is in the heading of the column in the table containing the observations.

Table 2.6 shows an example of this taken from an experiment in which the electrical resistance of a copper wire was measured as the temperature increased.

Table 2.6: Variation of electrical resistance with temperature of a copper wire

Temperature (°C) ± 0.5°C	Electrical resistance of length of copper wire (Ω) ± 0.001 Ω
8.0	0.208
16.5	0.213
23.5	0.222
32.0	0.229
40.5	0.232
54.5	0.243

There is more to be said about experimental uncertainties. If we have to combine measurements in order to calculate another quantity, how do uncertainties in the individual measurements add together to give an uncertainty in the calculated quantity? This and related matters are dealt with in chapter 4.

2.5 Significant figures

If a value obtained from a particular experiment is recorded as 6.12, this implies that the actual value lies between 6.11 and 6.13. If the value is written as 6.124, then this implies that the actual value lies between 6.123 and 6.125. Writing a value as 6.12 is to give it to three *significant figures*, and to write it as 6.124 is to give it to four significant figures. After making a measurement, the first inclination is to record a value which contains as many figures as the instrument provides. We should, wherever possible, assess and record the uncertainty in the measured value as discussed in the previous section (and more fully in chapter 4). Omitting this information obliges the reader to make an educated guess as to the size of the uncertainty based on the number of significant figures presented.

If the value 0.001 030 6 is recorded during an experiment, how many significant figures are implied by the way the number is written? We count the number of figures between the first non-zero figure and the last figure inclusive. In the case of the value 0.001 030 6, the first non-zero figure is a '1'. There are four remaining figures, making *five* significant figures in all.[3]

This rule for counting significant figures is fine except for situations in which we have whole numbers greater than ten that end with one or more zeros, for example 20, 1670, 3400, 545 000. Unless we are told explicitly that in the number 1670 all of the figures are significant, it is usual to infer that the number is known only to the nearest ten, and therefore the zero is *not* significant. It follows that 1670 is a number known to three significant figures. We will say more about this in section 2.5.3.

Example

How many significant figures appear in the following numbers?
(a) 1.654
(b) 0.004 37
(c) 64 000
(d) 1.20
(e) 0.100 007 38

Answers

(a) four
(b) three: the first non-zero figure (in this case the '4') is the first and most significant figure
(c) two
(d) three: if the zero is not significant then the number should have been written as 1.2
(e) eight

EXERCISE C

How many significant figures are implied by the way the following numbers are written?
(i) 3.24
(ii) 0.0023
(iii) 83 400
(iv) 1.010
(v) 10.5

---

3. The zeros that precede the '1' serve to place it in the correct position with respect to the decimal point, and are not regarded as significant figures.

2.5.1 Rounding numbers

When the result of a calculation produces a number that has many figures (which can easily happen if you use a pocket calculator), we may need to reduce the number of figures that appear. If, for example, the number 1.356 334 2 has to be reduced to two significant figures (we term this 'rounding' the number), a decision has to be made as to whether the second figure (in this case the 3) should be left as it is or increased by one. This is done by considering the third figure. If that figure is 5 or greater, the second figure is rounded up. If the third number is less than 5, the second figure is left alone. So, in this example, the number 1.356 334 2 to two significant figures would be 1.4.

In a calculation which involves a number of arithmetic steps, it is very good advice *not* to round numbers until all the calculations have been completed, otherwise the rounding process itself can have a large effect on the numbers that emerge from the calculations (see example in section 6.2.2).

EXERCISE D

Round the following numbers to three significant figures:
- (i) 18.92
- (ii) 0.107 59
- (iii) 725.4
- (iv) 1.7602
- (v) 62 654

2.5.2 Calculations and significant figures

Suppose you are required to find the cross-sectional area of a cylindrical rod after finding its diameter to be 8.9 mm. The relationship between the area of a circle, A, and its diameter, d, is:

$$A = \frac{\pi d^2}{4}$$

so that[4]

$$A = \frac{\pi(8.9 \text{ mm})^2}{4} = 62.211\ 388\ 52 \text{ mm}^2$$

The calculation was done using a pocket calculator capable of giving answers up to ten figures long. There is something disturbing here. The diameter d is known to two significant figures, and yet the area is given to ten. Is this reasonable? The answer is no. If we

---

4. When calculations involve quantities with units, we will include the units in any arithmetic steps shown.

had a computer that gave results of calculations up to a hundred figures long, we could have written A to that many figures, but what would they mean? The answer is, absolutely nothing!

If you are required, as in the example above, to perform a calculation in which the uncertainties in the quantities are not known, and all you have to work with is the number of significant figures in each quantity in the calculation, then the following rules are useful.

- **Rule 1:** When multiplying or dividing numbers: *identify the number in a calculation that is given to the **least** number of significant figures. Give the result of the calculation to the **same** number of significant figures.*

  For example: multiplying 3.7 by 3.01 gives 11.137. The number 3.7 has the least number of significant figures (two) and so we should give the answer as 11.

- **Rule 2:** When adding or subtracting numbers: *round the result of the calculation to the same number of decimal places as the number in the calculation given to the **least** number of decimal places.*

  For example: adding 11.24 and 13.1 gives 24.34. Using rule 2 we must give the result to one decimal place, that is, as 24.3.

EXERCISE E

Using the rules given in this section, write down the results of the following calculations to an appropriate number of significant figures:

(i) 1.2×8

(ii) 13.0×43.23

(iii) 0.0104×0.023

(iv) $33 + 435.5$

(v) $\dfrac{14.1}{76.3}$

(vi) $105.55 - 34.2$

2.5.3 Significant figures and scientific notation

It is not always clear how many figures in a number are significant. By changing the unit in which a number is expressed it can appear that the number of significant figures changes. For example, suppose in an experiment a particular time interval was recorded as 346 s. We could choose to write the time in other units such as milliseconds or microseconds. These would be written as 346 000 ms and 346 000 000 μs, respectively. In both cases the number of significant figures remains as three. However, if someone asked you for your value for the time interval to be expressed in ms, how would *they* know that of your value of 346 000 ms only the first three figures were significant? It *is* possible that you used a timing device capable of resolution of 1 ms and that the time interval came out to be, to the nearest millisecond, 346 000 ms, that is, *all* six figures are significant. The way to get around this difficulty is to present numbers in *scientific notation*.

In scientific notation the first non-zero figure that appears is followed by a decimal point, so that 346 becomes 3.46. To bring the number back to its original value we must multiply 3.46 by 100 or 10^2. So we can now write the time interval as 3.46×10^2 s. In terms of milliseconds and microseconds this becomes, 3.46×10^5 ms and 3.46×10^8 μs respectively.[5]

The number of significant figures is equal to the number of figures that appear to the left of the multiplication sign. In situations where a number lies between 1 and 10, for example 7.15, we could write this as 7.15×10^0. Though this is technically correct, it is more usual for the number to be expressed as 7.15.

Table 2.7 contains a variety of numbers and their representation in scientific notation (we assume here that all the figures given are significant).

Table 2.7: Examples of numbers expressed in scientific notation

Number	In scientific notation
12.65	1.265×10^1
0.000 23	2.3×10^{-4}
342.5	3.425×10^2
34 001	3.4001×10^4

EXERCISE F

1. Give the following numbers in scientific notation to *four* significant figures.
   (i) 0.005 654 2   (iv) 3 400 042
   (ii) 125.04   (v) 0.000 000 100 092
   (iii) 93 842 773
2. Repeat question (1.) but this time give the numbers to *two* significant figures.

---

5. Occasionally a large number such as 3.46×10^5 may be written as 3.46E5 (which means the same thing), or a small number such as 1.3×10^{-8} may be written as 1.3E−8. Computers commonly represent very large and very small numbers this way.

2.6 Orders of magnitude

When an experiment is performed, we often have a feel for the size of the numerical values that should emerge. This may come from having performed a similar experiment in the past, or just from everyday familiarity with the quantity being studied.

For example, if we were to measure the velocity of a car moving along a main street and found it to be 4×10^6 m s^{-1}, we should suspect something to be wrong. Equally, when measuring the mass of a glass beaker we should look again if that mass turned out to be 49 kg. We are saying that there are many situations in which it should be possible to assess how sensible the numbers are that emerge from an experiment, at least to within a factor of 10 of the 'actual' value. We speak of knowing the value to within an *order of magnitude*.

The habit of estimating the expected measured value to within an order of magnitude is very helpful for avoiding gross (and embarrassing) mistakes. Imagine you are given a voltmeter and battery that has been removed from a radio and are asked to determine the output voltage of the battery. People will be amused if you report that voltage as 1500 V, and may not take anything else you have to say seriously.

2.7 Comment

In this chapter we have considered a number of matters concerning the presentation of numerical data. No matter how faithfully we record our experimental values, taking due regard of things such as uncertainties and orders of magnitude, we still face a difficulty: absorbing all the data at once so that we can identify relationships between measured quantities is not easy if the data remain in tabular form. A most effective way of presenting the data in order to reveal relationships and anomalies is to plot a graph. We will deal with graph plotting in the next chapter.

PROBLEMS

1. The mass of a glass beaker was measured with a balance and found to be 45.64 g. When a liquid was added to the beaker, the balance indicated 92.5 g. Give the mass of the liquid in the beaker to the appropriate number of significant figures.

2. In a heat transfer experiment an amount of heat, Q, is transferred to silver at its melting point causing a mass, m, of the silver to melt. The relationship between m and Q is $Q = mL$ where L represents the heat of fusion of the silver.

   Given that $L = 88 \times 10^3$ J kg^{-1} and $Q = 4550$ J, calculate the mass of silver melted in kg. Express your answer in scientific notation to the appropriate number of significant figures.

3. As part of an experiment, a student was required to find the density, ρ, of a small solid metal sphere. Density, ρ, is given by $\rho = \dfrac{m}{V}$, where m is the mass of the sphere, and V is its volume ($V = \dfrac{4\pi r^3}{3}$, where r is the radius of the sphere). The student's notebook entry for this part of the experiment was as follows:

   Mass of sphere = 0.44 g

   Diameter of sphere = 4.76 mm.

   Using formula for volume of sphere
   $$V = \frac{4\pi \times (4.76)^3}{3} = 451.761\,761\,.$$

   Therefore density of sphere
   $$\rho = \frac{0.44}{451.761\,761} = 9.739\,647 \times 10^{-4}$$

   The student's notebook entry contains a number of mistakes or omissions. Can you name and, where possible, correct them?

GRAPHICAL PRESENTATION OF DATA

3

3.1 Overview: the importance of graphs

Our ability to take in information when it is presented in the form of a picture is so good that it seems natural to exploit this talent when analysing data obtained from an experiment. When data are presented pictorially, trends can be detected that we would be unlikely to recognise if the data were given only in tabular form. This is especially true in situations where a set of data consists of hundreds or thousands of values, which is a common occurrence when a computer is used to assist data gathering. Additionally, a pictorial representation of data in the form of a graph is an excellent way to summarise many of the important features of an experiment. A graph can indicate:

(i) the range of measurements made

(ii) the uncertainty in each measurement

(iii) the existence or absence of a trend in the data gathered. For example, the plotted points may lie in a straight line, a curve, or may appear to be scattered randomly across the graph paper.

(iv) which data points do *not* follow the general trend exhibited by the majority of data.

x-y graphs (also known as *Cartesian coordinate* graphs) are used extensively in science and engineering to present experimental data and it is those that we will concentrate upon in this chapter.

3.2 Plotting graphs

An *x-y* graph possesses horizontal and vertical axes, referred to as the *x-* and *y*-axis, respectively. Each data point plotted on the graph is specified by a pair of numbers termed the *coordinates* of the point. For example, point A in figure 3.1 has the coordinates $x = 20$, $y = 50$. The coordinates of the point may be written in shorthand as (x, y), which in the case of point A on figure 3.1 would be (20, 50). To assist in the accurate plotting of data points, graph paper may be used on which are drawn vertical and horizontal gridlines, as shown in figure 3.1.

The x-coordinate is sometimes referred to as the *abscissa* (or more loosely, but more commonly, the 'x-value' of a data point) and the y-coordinate as the *ordinate* (or the 'y-value' of a data point).

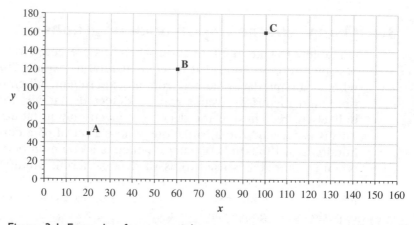

Figure 3.1: Example of an x-y graph

To be useful for displaying experimental data, a graph requires more informative axes labels than those shown in figure 3.1. Labels need to be attached to the axes which indicate the physical quantities being studied. Also required on the axes are the units in which the measurements were made. These and other matters are dealt with in the remaining sections of this chapter.

3.2.1 Dependent and independent variables

The quantity which is controlled or deliberately varied during an experiment is generally referred to as the *independent* variable and is plotted as the x-coordinate. The quantity that varies in response to changes in the independent variable is referred to as the *dependent* variable and is plotted as the y-coordinate. To take an example: if in an experiment we were to raise the temperature of an aluminium rod, we would find that its length increases. Here the temperature is the controlled quantity and is therefore the independent variable which is plotted as the x-coordinate. The length of the rod increases *as a consequence* of the temperature increase and is the dependent variable which is plotted as the y-coordinate. We say formally that the increase in the length of the rod is a *function* of temperature.

Table 3.1 shows values of the length of an aluminium rod as the temperature changes from 0°C to 250°C. These data are plotted on the x-y graph in figure 3.2.

Table 3.1: Length of an aluminium rod at various temperatures

Temperature (°C)	0	25	50	75	100	125
Length (m)	1.1155	1.1164	1.1170	1.1172	1.1180	1.1190

Temperature (°C)	150	175	200	225	250
Length (m)	1.1199	1.1210	1.1213	1.1223	1.1223

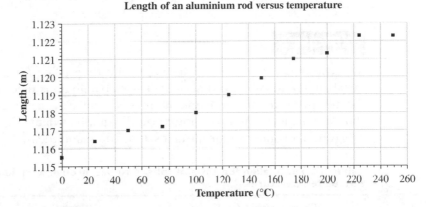

Figure 3.2: Graph of the variation of the length of an aluminium rod with temperature

3.2.2 Title, labels and units

The graph in figure 3.2 is typical of those used in science and engineering. It has:

(i) a title indicating the relationship being investigated. When it is stated that quantity 'A' (in this case the length) is plotted *versus* or *against* quantity 'B' (in this case the temperature), then quantity 'A' is plotted on the y-axis and quantity 'B' on the x-axis.

(ii) axes that are clearly labelled with the names of the quantities under study and their units of measurement. The unit of measurement is given in brackets after the name of the quantity represented on that axis. (It is possible to indicate the unit in another manner, as described in section 2.3, but I prefer to indicate units in brackets).

Occasionally you will encounter graphs in which the scale on the y-axis is proportional to the magnitude of the dependent quantity, but the absolute value of the quantity (that is, the quantity expressed in a recognised system of units) is not given. For example, in an experiment in which light intensity is measured as a function of position or time, we may want to know the *relative* change in intensity as the experiment proceeds. So long as an output of the light detector is proportional to the intensity of light, it is not necessary to convert that output to the SI unit for intensity (which, for luminous intensity, would be the candela). In such a case, the y-axis of the graph normally would be labelled *Light Intensity (arbitrary units)*.

EXERCISE A

An experiment is performed in which the time taken for a small metal sphere to fall a fixed distance through a liquid is recorded as the temperature of the liquid increases. The data gathered are shown in table 3.2 and are plotted on the graph in figure 3.3.

Figure 3.3 contains four mistakes. Can you name them?

Table 3.2: Time taken for a metal sphere to fall through a liquid at various temperatures

Temperature (°C)	Time (s)
21	62
26	48
30	35
37	26
42	22
46	19
51	17

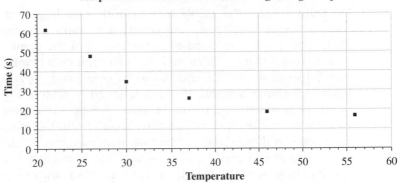

Figure 3.3: Graph indicating time taken for a metal sphere to fall through a liquid at various temperatures (with mistakes)

3.2.3 Scales and symbols

In the majority of situations it is sensible to choose scales so that the points that are plotted fill the available graph paper, leaving enough space to add labels, units and a title. When indicating data points on a graph, it is preferable to make the symbol representing a data point too big rather than too small. Figure 3.4 shows three sets of data, describing the motion of three cyclists, plotted on the same axes. The points indicated by the symbols ⊙ and ⊡ are easy to see, but those points indicated by small dots could be overlooked or mistaken for spurious marks on the paper.

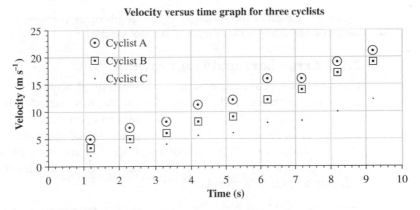

Figure 3.4: Graph of velocity as a function of time for three cyclists

3.2.4 Origins

On many graphs, the numbering of both axes begins at zero, so that the bottom left-hand corner of the graph has coordinates $(0, 0)$. This point is referred to as the *origin* of the graph. There is no rule to say that we *must* include the origin on a graph. To do so may cause important information to be concealed. As an example, figure 3.5 shows a graph of the data given in table 3.1 when the origin is included. The points in figure 3.5 lie almost horizontally and might lead us to conclude (incorrectly) that the length of the aluminium rod does not change with temperature. Including the origin has forced the use of a y-scale that is much too coarse to show clearly how the length of the rod varies with temperature. Figure 3.2, which shows the same data but does not include the origin, indicates much more clearly the relationship between length and temperature.

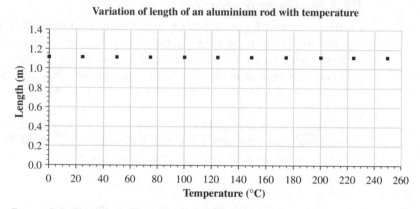

Figure 3.5: Graph of the variation of the length of an aluminium rod with temperature changes (with poor choice of y-scale)

3.2.5 Error bars and line drawing

Uncertainties in measurements were briefly discussed in section 2.4. It is possible to indicate the size of the uncertainties in the x- and y- quantities by attaching *error bars* to each data point on the graph.[1] Error bars are vertical and/or horizontal lines that extend from a data point. The length of each line is a measure of the size of the uncertainty in the quantity. Let us illustrate this by considering a specific example.

---

1. We could say 'uncertainty bars' instead of error bars, but the latter term is much more commonly used.

Table 3.3 shows data from an experiment in which the temperature of an object has been recorded as it cooled. In the heading of each column there is an estimate of the uncertainty in the measured quantity. In this example, time is plotted as the *x*-quantity, and temperature as the *y*-quantity.

Table 3.3: Variation of temperature with time for a hot object

Time (s) ±5 s	Temperature (°C) ±4°C
10	125
70	116
125	104
190	94
260	87
320	76
370	72

To indicate the uncertainty in the time, we draw horizontal error bars as follows:

To indicate the uncertainty in the temperature, we draw vertical error bars as follows:

Figure 3.6 shows both the vertical and horizontal error bars attached to all the data points. In this example, the size of the uncertainties do not vary from one point to the next. In many experiments the uncertainties are not constant from one measurement to the next, so that the size of the error bars vary from point to point on the graph. Where error bars are too small to plot clearly, it is advisable to omit them.

Assuming that the *y*-quantity varies smoothly with changes in the *x*-quantity, we can draw a line through the points as shown in figure 3.6. In situations where points lie along a curve, it can be difficult to draw a line 'by hand' that passes close to all the points. A strip of stiffened rubber (sometimes referred to as a *flexi-curve*) can assist greatly when drawing curves on graphs. It is difficult to extract reliable quantitative information about the relationship between *x* and *y* from a curve through the points, the line acting more as a 'guide to the eye'. If quantitative information *is* sought, it is better (if possible) to 'linearise' the data so that a straight line is produced. Straight-line graphs are dealt with in section 3.3 and linearisation in section 3.3.5.

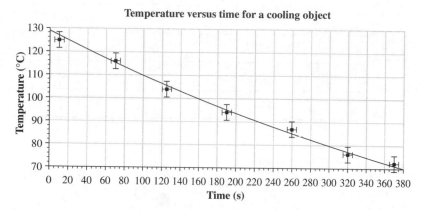
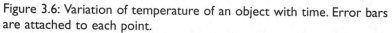

Figure 3.6: Variation of temperature of an object with time. Error bars are attached to each point.

3.2.6 When to plot a graph

A graph may be plotted either as an experiment proceeds or after the measurements have been completed. Plotting graphs during an experiment can be an efficient use of time as well as allowing for the immediate identification of interesting or unexpected features in the data.

However, plotting a graph during an experiment has its disadvantages:

(i) A knowledge is required of the likely range of *x*- and *y*-values, otherwise scales cannot be drawn in advance. Such knowledge is unlikely to be available where an experiment is being performed for the first time.

(ii) The experiment may require careful and continuous attention. The break in concentration needed to plot a data point may cause an important effect to go unnoticed or a critical adjustment to be delayed.

In many situations it is advisable to attend to the problem of ensuring that measurements are made as carefully as possible, leaving graph plotting until the end of the data-gathering stage of the experiment.

Once the data points are plotted they may appear to lie along a curve, a straight line or show no trend at all. We will look in detail at what is perhaps the most important type of graph: the linear (also known as the straight line) *x-y* graph.

3.3 Linear x-y graphs

Figure 3.7 shows data obtained from an experiment in which the magnetic field at a point close to a coil of wire (referred to as coil #1) was measured as the electrical current through the wire was increased.

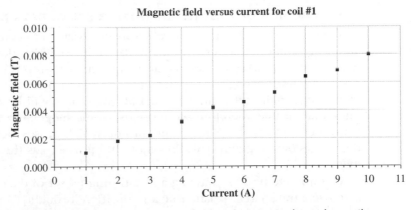

Figure 3.7: Variation of magnetic field with current through a coil

The experimental observations reveal that as the current increases through the coil, the magnetic field at a point near to the coil increases in direct proportion. This is an example of a linear *x-y* graph. We see that points on the graph do not lie exactly along a straight line. This is likely to be due to the experimental uncertainty that exists in the measurements that were made. Nevertheless, it would be reasonable to conclude that there is a linear relationship between magnetic field and current.

31

A useful tip when trying to decide quickly whether plotted data follow a linear path is to hold the graph paper up to your eyes and tilt the paper so that you can look along the line of the data points. It should be possible to discern quite easily whether the data follow a straight line or a line with slight but consistent curvature.

Linear graphs have an important place in the analysis of experimental data for the following reasons:

(i) Two important constants can be calculated from a straight line through the points. These are the gradient of the line and the point at which the line crosses the y-axis. These will be discussed in section 3.3.1 and again in chapter 6.

(ii) Departure from linearity, say at the extremes of the plotted data, can be observed.

(iii) Points that lie far from the line (often referred to as *outliers*) can be identified.

(iv) We can reliably predict values for the y-quantity for a chosen value of the x-quantity. Likewise, for a particular y-value we can use the straight line to find the corresponding value of x.

Let us deal with each of these matters in turn.

3.3.1 The line of 'best fit' through a set of data points

If we are satisfied that a linear relationship exists between the x- and y-quantities, it is very useful to be able to write down an equation that represents that relationship. The first step is to draw a straight line through the points on the graph. Due to experimental uncertainty, it is unlikely that all the points will lie exactly on a straight line, therefore we must use some judgement to estimate the position of the 'best' straight line which passes closest to the points on the graph. This line is often referred to as the line of 'best fit'.

A simple method for drawing a line through a set of data points is to use a transparent plastic rule and position it roughly along the line of the points. The next step is to move the rule until the data points on the graph appear to be scattered evenly above and below the line. It is recommended that the best line be drawn using a sharp pencil so that any mistake made in positioning the rule can be corrected.

Figure 3.8 repeats the data shown in figure 3.7, but includes two attempts at drawing the best straight line through the points.

Line 1 on the graph is a good effort at the best straight line. Looking closely, we see that five data points lie above the line, four below, and one point (where the current is 4 A) appears to be on the line. Not only are there approximately as many data points above the line as below, but as we move from left to right across the graph we see that data points are scattered about the line. A random distribution of points above and below the line is a distinguishing feature of the best straight line through a set of points.

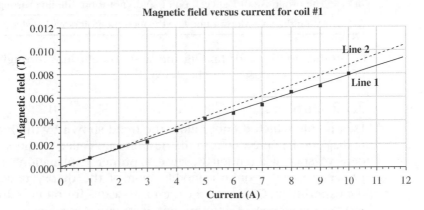

Figure 3.8: A linear *x-y* graph with two 'best' lines drawn through the points

Line 2 on the graph shown in figure 3.8 is less satisfactory than line 1 as the best line through the points. The line fits the first few points quite well, but at large values of current, it lies consistently *above* the data points. It also appears that the line has been deliberately drawn through the origin. All data points should be given equal weight, unless we have some good reason to do otherwise.[2] Consequently, even if the origin *is* a data point, no special effort should be made to force the line through it. If it is believed that the magnetic field *should* be zero when the current is zero, then we need to ask why the best line does not go through the origin. It is possible that the instrument measuring the magnetic field has a zero offset which was not taken into consideration,[3] or perhaps there was another source of magnetic field that was not accounted for. Whatever the situation, it is unwise to force a line to pass through the origin without a strong reason to support your action.

Table 3.4 summarises the steps to take when drawing a best straight line 'by eye'.

---

2. We will deal with situations where experimental data should be 'weighted' in chapter 6.
3. Offsets are discussed in section 4.4.1.

Table 3.4: How to draw the best line through a set of x-y data

Step	Action
(i)	Position a clear plastic rule along the plotted data points.
(ii)	Move the rule until the points are scattered as equally as possible above and below the line.
(iii)	The origin is not a special point so do not force the line through it.
(iv)	Using the rule, draw a fine line through the points with a sharp pencil.

Another method for finding the best straight line through a set of x-y data is discussed in chapter 6.

3.3.2 Outliers

Data points which do not follow the trend shown by the majority of data and lie far away from the best line are called 'outliers'. They deserve special attention because a proper explanation of their presence might require a modification of the theory underlying the experiment. Other, more common, reasons for outliers are:

(i) the existence of intermittent instrument problems such as electrical interference affecting measurements

(ii) incorrect recording of measured values during an experiment

(iii) a mistake when plotting the data points.

Unless the cause of an outlier can be found immediately, the best advice is to repeat the measurement to determine whether the outlying point is repeatable. To throw away a data point because it spoils the look of the graph could be to discard the most important measurement!

3.3.3 Interpolation and extrapolation

Once the best line has been drawn through a set of x-y data, it is an easy matter to find a y-value at any given x-value, or vice versa. If we find a value of y at an x-value that lies *within* the range of data points we have plotted, this is termed interpolation. For example, with reference to the graph in figure 3.8, we see that when the current is 2.5 A, the value of magnetic field that corresponds to that current found using the best line (line 1) is 0.0020 T.

When an x-value lies outside the measurement range, the corresponding y-value is found by extrapolation. Referring again to figure 3.8, when the current is 10.5 A, the corresponding magnetic field value is 0.0082 T.

Caution should be exercised when interpolating and even more so when extrapolating. It may be that the linear relationship breaks down at some value of x beyond the chosen measurement range, so that extrapolation is not valid. Though less likely, some non-linearity could exist *between* consecutive data points, especially if those points are widely separated. If non-linearity *is* suspected then it is sensible to repeat the experiment so that it covers both a greater number and wider range of x-values.

EXERCISE B

1. Using line 1 in figure 3.8, find the value of a magnetic field when the current is
   (i) 5.5 A
   (ii) 11.5 A
2. A hydrometer is an instrument which indicates the density of a liquid relative to that of pure water.[4] Table 3.5 shows measurements made with a hydrometer for water samples that contain various concentrations of salt.

Table 3.5: Variation of relative density with salt concentration

Salt concentration ($mg\ cm^{-3}$)	Relative density (no units)
0	1.005
50	1.034
100	1.066
150	1.095
200	1.122
250	1.150

   (i) Plot a graph of relative density versus salt concentration.
   (ii) Draw the line of best fit through the data.
   (iii) The relative density of a sample of water taken from Sydney Harbour is measured using the hydrometer and is found to have a value of 1.053. Use the line on your graph to estimate the salt concentration in the water from the harbour.

---

4. The scale on a hydrometer gives the *relative density* (r.d.) of a liquid which is defined as, r.d. = density of liquid/density of pure water. Being the ratio of two quantities that have the same unit, the relative density itself is a number without units.

3.3.4 The gradient and intercept of the best straight line

The y-coordinate of any point on a straight line can be related to the corresponding x-coordinate by the equation:[5]

$$y = mx + c \qquad (3.1)$$

The symbol m is termed the *gradient* of the line and c the *intercept*. An accurate determination of m and c is important as they are parameters that can be compared between various experimenters who are studying the relationship between the same physical quantities. Often, the value of an important physical parameter can be calculated using the values of m and c. Let us consider a specific example.

The amount that a material expands or contracts when its temperature changes is of great importance in the design of structures (for example, a bridge) as expansion that hasn't been allowed for could cause a structure to bend or buckle. In this context, it is important to know the *coefficient of linear expansion, α*, of the material as this permits the length increase of a material to be predicted given its original length and the temperature rise. If we were to use aluminium as a construction material, we could find α for aluminium by using the graph shown in figure 3.2. It turns out that (after the best line has been drawn through the points) α is equal to the gradient of the line divided by the intercept (see example 1 in section 3.3.5. for more details).

To understand how we can calculate m and c for a straight line, consider figure 3.9. Two points have been chosen on the line and their coordinates written as (x_1, y_1) and (x_2, y_2).

Using equation 3.1 we have:

$$y_1 = mx_1 + c \qquad (3.2)$$

$$y_2 = mx_2 + c \qquad (3.3)$$

Solving equations 3.2 and 3.3 to find m gives:

$$m = \frac{y_2 - y_1}{x_2 - x_1} \qquad (3.4)$$

$y_2 - y_1$ is sometimes referred to as the *rise* and $x_2 - x_1$ as the *run*, so another way of writing the gradient is:

$$m = \frac{\text{rise}}{\text{run}} \qquad (3.5)$$

5. It is quite common to see this equation written using other symbols for the gradient and intercept, such as $y = mx + b$, or $y = ax + b$, or $y = bx + a$.

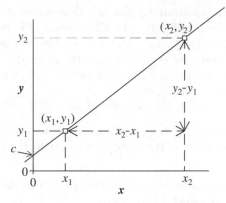

Figure 3.9: Straight line through points on an x-y graph

The two points indicated by crosses in figure 3.10 have coordinates, $x_1 = 1, y_1 = 4, x_2 = 11$ and $y_2 = 25$. Applying equation 3.4, we find that m is given by:

$$m = \frac{25-4}{11-1} = 2.1$$

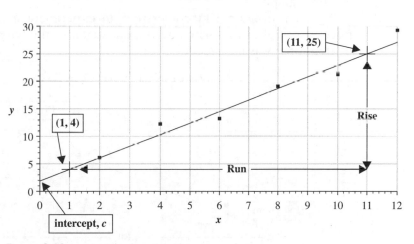

Figure 3.10: x-y graph showing the run and rise

When finding the gradient you must:
(i) only choose points that lie *on* the best line. This normally means that data points *cannot* be used for determination of the gradient.
(ii) choose points on the line that are well separated, as in figure 3.10, as this lessens the effect of any inaccuracy arising from measuring the run and rise from the graph.

37

Using equation 3.1, we see that when $x = 0$, then $y = c$. For the line shown in figure 3.11, the intercept is equal to 2.

In situations where the x-axis does not begin at $x = 0$, the value of c cannot be obtained directly from the graph. In such cases c can be calculated by first finding m as above, and then rearranging equation 3.1 so that:

$$c = y - mx \qquad (3.6)$$

By choosing any point on the line, and substituting the x- and y-coordinates of that point into equation 3.6 we can find the value of c.

3.3.4.1 Gradient and intercept when dealing with experimental data

There are situations, particularly in mathematics, where the points on a graph do not represent experimental data and so the axes have no units of measurement associated with them. By contrast, the majority of graphs in science and engineering show relationships between physical quantities and so *do* have units[6]. The units must be included when calculating the gradient and intercept of a line on a graph.

The data in figure 3.11 come from an experiment in which a force was applied to a wire and the extension of the wire caused by that force was measured.

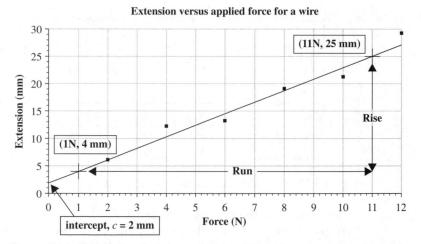

Extension versus applied force for a wire

Figure 3.11: Graph showing the variation of the extension of a wire with applied force

---

6. There are occasions when ratios of two quantities, each with the same unit, are plotted on the x- or y-axis. Such ratios have no units.

The intercept has the same units as the quantity on the y-axis. In the case of the graph in figure 3.11, the intercept is 2 mm. The gradient also has units which are found by referring back to equation 3.4. As well as inserting the coordinates into the equation, we also include the units, so that:

$$m = \frac{(25 - 4) \ mm}{(11 - 1) \ N} = \frac{21 \ mm}{10 \ N} = 2.1 \ mm \ N^{-1}$$

EXERCISE C

The electrical resistance of a specimen of iron is measured over a temperature range 40°C to 90°C. Table 3.6 shows data gathered in the experiment.

Table 3.6: Variation of resistance with temperature for a specimen of iron

Temperature (°C)	Resistance (Ω)
40	6.22
50	6.51
60	6.74
70	7.05
80	7.29
90	7.51

Using the data above, plot a graph of resistance versus temperature, beginning the x-axis at 40°C and the y-axis at 6 Ω Draw the best line through the points and use equation 3.6 to find the resistance of the specimen at 0°C. What assumption is made when using this method to find the resistance at 0°C?

3.3.4.2 Uncertainties in gradient and intercept

The reason for drawing a straight line as close as possible to the data points is to obtain the best possible estimate of the *quantitative* relationship between the quantities plotted on each axis. As there is an uncertainty in each measurement, there must also be an uncertainty in the gradient and intercept of the line drawn through the points. If error bars have been attached to each point, we can use these to assist in estimating the uncertainty in the gradient and intercept. Figure 3.12 shows data gathered in a crystal-growing experiment in which the length of the crystal is plotted against time of growth. Error bars have been attached to each point. As the uncertainty in time is negligible in this experiment, the horizontal error bars are omitted.

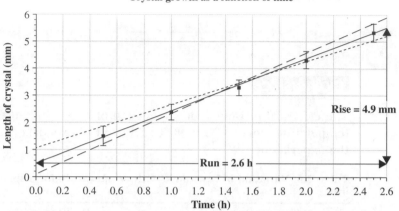

Figure 3.12: Variation of the length of a growing crystal with time

In order to estimate the uncertainty in the gradient and intercept we draw three lines through the data. The first is the best straight line which passes as close as possible to all the points, as discussed in section 3.3.1. The gradient of this line is:[7]

$$\frac{4.9 \text{ mm}}{2.6 \text{ h}} = 1.9 \text{ mm h}^{-1}$$

The other two lines are drawn so that they give the maximum and minimum gradient consistent with the error bars. So, for example, the line with the maximum gradient is drawn so that it passes through *all* the error bars, but for the data on the extreme right the line passes through the top of the error bars, and for the points at the extreme left the line passes through the bottom of the error bars. The gradient of this steepest line is:

$$\frac{5.7 \text{ mm}}{2.6 \text{ h}} = 2.2 \text{ mm h}^{-1}$$

The minimum gradient is found from the line which passes through the bottom of the error bars for the data points on the

---

7. The run and rise are shown for the best straight line in figure 3.12. In the interests of clarity, the run and rise for the other two lines have been omitted from the graph.

extreme right, and the top of the error bars for the data on the extreme left. The gradient of this line is:

$$\frac{4.1 \text{ mm}}{2.6 \text{ h}} = 1.6 \text{ mm h}^{-1}$$

We can now write the gradient and the associated uncertainty as (1.9 ± 0.3) mm h^{-1}.

In order to obtain the uncertainty in the intercept we locate where the three lines cross the y-axis. From the graph, the best value for the intercept is 0.5 mm with an upper value of 1.1 mm and a lower value of 0.1 mm. Now we can give the intercept and estimate of the uncertainty in the intercept (found by dividing (upper value − lower value) by 2) as:

$$c = (0.5 \pm 0.5) \text{ mm}$$

Another method for finding the uncertainties in the gradient and intercept is discussed in section 6.2.4.

3.3.5 Linearising equations

In many experiments we have some prior knowledge about the relationship between the quantities being studied. If this is the case, and if the relationship can be expressed as an equation, then we can often choose what to plot on each axis to produce a straight-line graph. We speak of *linearising* the equation. Let us consider a specific example.

In an experiment, a body of mass, M, is attached to the end of a spring and set oscillating vertically. The period of the body is measured as the mass of the body is increased. The graph in figure 3.13 shows how the period varies with mass.

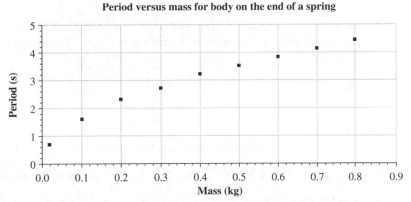

Figure 3.13: Variation of the period of an oscillating body with the mass of the body

Looking along the points in figure 3.13, it is clear that the relationship between period and mass is not linear. If, instead of plotting period versus mass, we plot period versus $(mass)^{\frac{1}{2}}$, the outcome is the straight-line graph shown in figure 3.14.

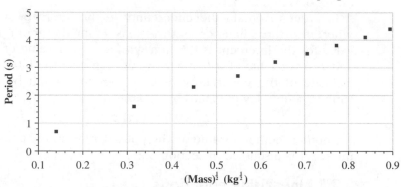

Period versus $(mass)^{\frac{1}{2}}$ for a body on the end of a spring

Figure 3.14: Period of a body attached to a spring plotted against $(mass)^{\frac{1}{2}}$

The graph was linearised by changing the quantity plotted on the x-axis. Why did plotting $(mass)^{\frac{1}{2}}$ on the x-axis linearise the graph? To understand this we must look again at the equation of a straight line, $y = mx + c$, and relate this to the equation which represents the relationship being studied.

The equation which relates the period of oscillation, T, to the mass, M, on the end of a spring is:

$$T = 2\pi\sqrt{\frac{M}{k}}$$

where k is a constant. To make comparison with $y = mx + c$ easier, the equation can be rewritten with $y = mx + c$ below, as follows:

$$T = \frac{2\pi}{\sqrt{k}}\sqrt{M} \qquad (3.7)$$

$$y = m\ x\ +\ c$$

We see that if T is plotted on the y-axis and $\sqrt{M}$ is plotted on the x-axis, the gradient of the line is $\dfrac{2\pi}{\sqrt{k}}$. As there is no constant appearing as a separate term on the right-hand side of equation 3.7, the intercept is equal to zero.

There are a large number of situations where linearisation is possible, but the process of arranging an equation into the form, $y = mx + c$, does take practice so we will consider three representative examples.

Example 1

In an experiment to study the variation of length with temperature for a metal rod, the equation relating length, l, to temperature, θ, is:

$$l_\theta = l_0 (1 + \alpha\theta) \qquad (3.8)$$

l_θ is the length of the rod at the temperature $\theta°C$, l_0 is the length at $0°C$, and α is the temperature coefficient of linear expansion.

In order to write equation 3.8 in the form $y = mx + c$, the first step is to expand the right-hand side of the equation by multiplying through by l_0, giving:

$$l_\theta = l_0 + l_0\alpha\theta$$

Now we compare this equation with $y = mx + c$. As we have done before, we will write $y = mx + c$ below the equation and group together corresponding quantities as shown below.

$$l_\theta = l_0\alpha\ \theta + l_0$$
$$y\ \ =\ \ m\ x\ +\ c$$

In this example, plotting l_θ versus θ should give a straight line with a gradient equal to $l_0\alpha$ and intercept, l_0.

Example 2

The displacement of a body, s, accelerating uniformly with acceleration, a, is measured at various times, t. The relationship between s and t is:

$$s = ut + \tfrac{1}{2}at^2 \qquad (3.9)$$

where u is the initial velocity of the body.

In order to linearise equation 3.9, divide throughout by t then rearrange, to give:

$$\frac{s}{t} = \tfrac{1}{2}at + u \qquad (3.10)$$

Proceeding as before, we can compare the quantities in equation 3.10 with $y = mx + c$:

$$\frac{s}{t} = \tfrac{1}{2}a\ t + u$$
$$y\ =\ m\ x\ +\ c$$

Plotting $\dfrac{s}{t}$ versus t produces a straight line with gradient $\frac{1}{2}a$ and intercept u.

This example differs from the previous one in that the independent variable (in this case the time, t) appears on both sides of the equation. In most cases, the independent variable appears only on the right-hand side of the equation. However, some situations (such as this one) require that this rule be relaxed for linearisation to occur.

Example 3

The equation:

$$N = N_0 e^{(-\lambda t)} \tag{3.11}$$

relates the number of undecayed nuclei, N, that remain in a sample of radioactive material, after a time t has elapsed. N is the dependent variable, t the independent variable. N_0 and λ are constants.

To linearise an equation which contains a number raised to a power, the first step is to take the logarithms of both sides of the equation. In this example, the number raised to the power $-\lambda t$ is the constant, e (which has the value, to seven significant figures, of 2.718 282). If we take logarithms to the base e of both sides of equation 3.11 and rearrange, we obtain:

$$
\underbrace{\ln (N)}_{y} = \underbrace{-\lambda}_{m}\, \underbrace{t}_{x} + \underbrace{\ln (N_0)}_{c}
$$

Plotting $\ln (N)$ versus t produces a straight line with gradient equal to $-\lambda$, and intercept equal to $\ln (N_0)$.

EXERCISE D

Table 3.7 contains equations taken from various fields in science and engineering. For each equation, the dependent variable, the independent variable and the constants in the equation are indicated.

For each of the equations in table 3.7, answer the following questions:

(i) What would you plot in order to obtain a straight line?

(ii) How are the gradient and intercept related to the constants in the equation?

44

Table 3.7: Equations to be arranged in the form y = mx + c

Equation number	Equation	Dependent variable	Independent variable	Constant(s)
1	$F = \mu N$	F	N	μ
2	$v = u + at$	v	t	u, a
3	$R = AT + BT^2$	R	T	A, B
4	$I = I_0\exp(-\mu x)$	I	x	I_0, μ
5	$T = 2\pi\sqrt{\dfrac{l}{g}}$	T	l	$2\pi, g$
6	$\dfrac{1}{u} + \dfrac{1}{v} = \dfrac{1}{f}$	v	u	f
7	$H = C(T - T_0)$	H	T	C, T_0
8	$I = AV\exp(-BV^2)$	I	V	A, B
9	$T_w = T_c - kR^2$	T_w	R	T_c, k

3.4 Logarithmic graphs

The graphs we have considered so far have been plotted using axes with scales that increase linearly. There are circumstances, however, where no matter how careful we are when choosing scales, useful information is obscured. As an example of this, consider table 3.8 which shows data obtained from a study of the electrical characteristics of a silicon diode.

Table 3.8: Current variation with applied voltage for a silicon diode

Voltage (V)	Current (A)
0.35	9.0×10^{-7}
0.40	3.0×10^{-6}
0.45	5.0×10^{-5}
0.50	2.0×10^{-4}
0.55	1.7×10^{-3}
0.60	1.5×10^{-2}
0.65	7.5×10^{-2}
0.70	0.55
0.75	3.5

A graph of the data in table 3.8, plotted using linear scales, is shown in figure 3.15.

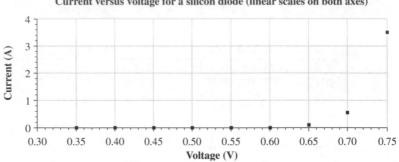

Figure 3.15: Current variation with voltage for a silicon diode: linear x-y scales chosen

The current data shown in figure 3.15 span so many orders of magnitude that, in choosing a scale to fit the large values of current on the graph, the values of currents measured at voltages less than 0.6 V cannot be read from the graph. Data spanning many orders of magnitude can be better accommodated using graph paper in which the scales do not increase linearly, but *logarithmically*. Figure 3.16 shows the data from table 3.8, but, in this instance, equal distances on the vertical scale correspond to changes by *powers* of ten.

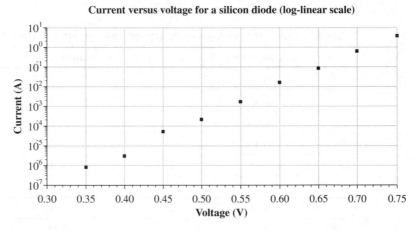
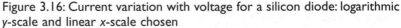

Figure 3.16: Current variation with voltage for a silicon diode: logarithmic y-scale and linear x-scale chosen

Figure 3.16 shows the current versus voltage data plotted on a log-linear scale[8]. By that we mean that one axis increases logarithmically (in this example the y-axis), and the other increases linearly.

Although in this example, plotting the quantities on a log-linear scale has produced points which lie in a straight line (indicating that there is a logarithmic relationship between current and voltage), in general, a straight line cannot be expected.

In order to plot the current-voltage data shown in table 3.8, it is necessary to have log-linear paper in which the y-axis spans eight powers of ten. This represents a difficulty when using commercially available paper with logarithmic scales. It is common to find scales spanning two, three or four powers of ten (often referred to as two- three- or four-cycle 'log' paper). To cover eight orders of magnitude, it is necessary to combine two pieces of four-cycle paper. This can be difficult and time consuming.

Another way to approach the plotting of data that covers many orders of magnitude is to take the logarithms of that data and plot them on conventional linear graph paper.[9] It does not matter to which base we take the logarithms, but by far the most common method is to take logarithms to the base e or to the base ten.

Table 3.9 shows the current and voltage data for the diode given in table 3.8 with the addition of a column giving the logarithm of the current values to the base ten. A plot of $\log_{10}$ (current) versus voltage is shown in figure 3.17 overleaf.

Table 3.9: Current-voltage data for a silicon diode

Voltage (V)	Current (A)	$\log_{10}$ (current)
0.35	9.0×10^{-7}	-6.046
0.40	3.0×10^{-6}	5.523
0.45	5.0×10^{-5}	-4.301
0.50	2.0×10^{-4}	-3.699
0.55	1.7×10^{-3}	-2.770
0.60	1.5×10^{-2}	-1.824
0.65	7.5×10^{-2}	-1.125
0.70	0.55	-0.260
0.75	3.5	0.544

8. This is sometimes referred to as a *semi-log* scale.
9. Often, the convenience and availability of linear graph paper makes this method of plotting data more attractive than using log-linear paper.

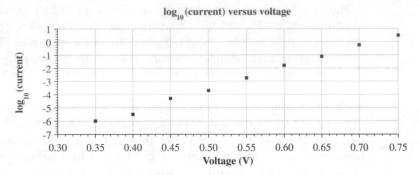

Figure 3.17: Log_{10} (current) as a function of applied voltage for a silicon diode

3.4.1 Log-log graphs

When both the x- and y-data span many orders of magnitude, then a log-log graph can be used to display the data. Another circumstance in which log-log graphs are useful is when there is believed to be a 'power law' relationship between the two quantities. An example of a power law relationship is:

$$I = Ad^n$$

where d is the independent variable, I is the dependent variable, and A and n are constants. If we take logarithms to the base ten of both sides of the equation, we get:

$$\log_{10}(I) = \log_{10}(A) + n \log_{10}(d)$$

Plotting $\log_{10}(I)$ versus $\log_{10}(d)$ will produce a straight line with intercept, $\log_{10}(A)$, and gradient, n. Plotting the data on log-log graph paper, which has scales which increase logarithmically along *both* the x and y-axes will also produce a straight line.[10]

EXERCISE E

Table 3.10 gives data obtained from an experiment in which the amount of energy emitted from a hot body each second, H, was recorded as the temperature of the body, T, increased. It is believed that there is a power law relationship between H and T of the form:

$$H = AT^n$$

---

10. Log-log graph paper is available for plotting data covering many orders of magnitude, but just as for log-linear graphs, it is often more convenient to use linear paper after taking the logs of the quantities.

Plot a log-log graph using the data in the table and draw the best straight line through the data. Use the best line to find the values of A and n.

Table 3.10: Energy emitted each second from a hot body as the temperature of the body increases

Temperature (K)	H (W)
1500	150
1600	190
1700	230
1800	300
1900	360
2000	440
2100	560
2200	680
2300	800
2400	930
2500	1100

3.5 Comment

Images are a powerful way to communicate information. Representing data from an experiment in the form of an x-y graph allows us to discuss relationships, assess scatter in data and permit the rapid identification of unusual or irregular features. A well laid out graph containing all of the components discussed in this chapter can act as a 'one-stop' summary of a whole experiment. Someone studying an account of an experiment will very often look at the graph(s) included in the account first to get an overall picture of the outcome of an experiment. The importance of graphs, therefore, cannot be overstated as they so often play a central role in the presentation and analysis of data.

PROBLEMS

1. In an experiment to study the frictional forces between bodies, a force was applied to a block of wood situated on a flat metal surface, as shown in figure 3.18 overleaf.

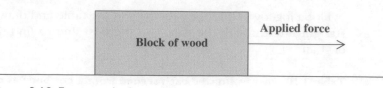

Figure 3.18: Force applied to a block of wood

It was observed that the minimum force required to cause the block to move increased as extra masses were attached to the block. Table 3.11 shows the relationship between the mass of the block and the minimum force required to move the block.

Table 3.11: Minimum applied force required to move the block as the mass of the block increases

Mass (kg) (± 0.01 kg)	Minimum force (N) (± 0.2 N)
0.52	3.1
0.58	3.6
0.64	3.9
0.75	4.4
0.88	5.2
1.01	6.2
1.21	7.4

(i) Plot a graph of minimum force against mass. Attach error bars to each point.

(ii) Draw the best line through the points and calculate the gradient and intercept of the line. Write the equation of the best straight line in the form, $y = mx + c$.

(iii) Using the equation of the best straight line, estimate the minimum force required to move the block, when the mass of the block is:
(a) 0.70 kg
(b) 1.30 kg

(iv) With the aid of the error bars, estimate the uncertainty in the gradient and intercept.

2. Table 3.12 shows the pressure, P, of a fixed volume of gas which was measured as the temperature, T, of the gas increased.

Table 3.12: Pressure of a fixed volume of gas as a function of temperature

P ($\times 10^5$ Pa)	T (K)
0.92	280
0.98	295
1.02	305
1.05	320
1.15	350
1.20	370
1.30	390

Plot a graph of pressure versus temperature and draw the best straight line through the points. Find:
(i) the gradient of the line
(ii) the intercept of the line (that is, the value of pressure corresponding to $T = 0$ K)
(iii) the pressure at 330 K and 400 K
(iv) the temperature corresponding to a pressure of 1.00×10^5 Pa.

3. An experiment was performed to determine the amount of mercury in river water using an analytical technique called 'atomic absorption spectrophotometry'. Calibration data of light absorption as a function of mercury concentration are shown in table 3.13.
(i) Plot a graph of absorption versus concentration.
(ii) Draw the best line through the points.
(iii) A sample of river water shows an absorption of 25 ± 3 arbitrary units. Use the best line on your graph to estimate the concentration of mercury in the river water and the uncertainty in the concentration.

Table 3.13: Light absorption as a function of mercury concentration

Mercury concentration (parts per billion)	Absorption (arbitrary units)
0.5	10
1.0	20
2.0	38
3.0	56
4.0	75

4. The electrical current, I, flowing around a circuit was measured as the resistance in the circuit, R, was varied. The relationship between I and R is given by:

$$I = \frac{E}{R + r}$$

where E is the e.m.f. of a battery in the circuit and r is its internal resistance (E and r are constants).

   (i) Linearise the equation above and state what you would plot on the x- and y-axes in order to obtain a straight line.

   (ii) State clearly how E and r are related to the gradient and/or the intercept of the graph.

5. At low temperature a ceramic conductor was observed to exhibit unusual electrical characteristics. Table 3.14 shows the variation of electrical resistance of a sample of the ceramic as the current through it increased.

Table 3.14: Resistance of a material as a function of current through the material

I (A)	R (Ω)
1.0×10^{-3}	6.0×10^{-4}
2.0×10^{-3}	2.2×10^{-3}
4.0×10^{-3}	6.3×10^{-3}
8.0×10^{-3}	2.0×10^{-2}
1.6×10^{-2}	4.2×10^{-2}
3.2×10^{-2}	1.2×10^{-1}
6.4×10^{-2}	3.4×10^{-1}
1.3×10^{-1}	1.1
2.6×10^{-1}	3.2
5.2×10^{-1}	9.5

Assume that the relationship between R and I can be written:

$$R = kI^n$$

where k and n are constants.

   (i) Plot a suitably linearised graph of the data given in table 3.14.

   (ii) Draw the best straight line through the points.

   (iii) Calculate the gradient and intercept of the line.

   (iv) Use the gradient and intercept to find values for k and n.

6. When sodium thiosulphate ($Na_2S_2O_3$) is mixed with hydro-chloric acid, a reaction occurs during which a sulphur precipitate is formed. An experiment was performed to study the rate of reaction as a function of concentration of $Na_2S_2O_3$. Table 3.15 shows the time for the reaction to reach a specific point as the concentration of $Na_2S_2O_3$ was increased.

Table 3.15: Time for reaction between $Na_2S_2O_3$ and hydrochloric acid as a function of concentration of $Na_2S_2O_3$

Concentration (M)	Reaction time, t (s)
0.20	25
0.16	30
0.14	32
0.12	40
0.08	56
0.02	116

It is believed that the reaction rate for these chemicals should depend linearly on the concentration of the $Na_2S_2O_3$ present.

(i) Taking the reaction **rate** to be equal to $\dfrac{1}{\text{reaction time}}$, draw up a table of reaction rate as a function of concentration.

(ii) Plot reaction rate as a function of concentration of $Na_2S_2O_3$.

(iii) Draw the best line through the points and calculate the gradient of the line.

DEALING WITH UNCERTAINTIES

4.1 Overview: what are uncertainties?

An experiment may consist of measurements made with simple equipment such as a stopwatch or a metre rule, or require highly sophisticated equipment such as that carried by satellites to monitor the hole in the ozone layer. However, all experiments have at least one thing in common: each measurement that is made is subject to experimental *uncertainty*. By that we mean that if we were to make repeated measurements of a particular quantity, we are likely to find a variation in the observed values. Although it may be possible to *reduce* an uncertainty by improved experimental method or the careful use of statistical techniques, it can never be eliminated.

In this chapter we will look at ways of recognising and dealing with experimental uncertainties.

4.1.1 Introductory example of experimental uncertainty

Consider an experiment in which a small object falls through a fixed distance and the time for it to fall is measured using a stopwatch. Table 4.1 shows ten values recorded for the time of fall.

Table 4.1: Times for object to fall a fixed distance

Time (s)	0.64 0.61 0.63 0.53 0.59 0.65 0.60 0.61 0.64 0.71

We might have hoped that on each occasion we measured the time of fall of the object, we would have obtained the same value. Unfortunately, this is not true in the case of the above experiment, and in general it is untrue of *any* experiment.[1] We must

---

1. There is an exception to this. When items are counted (for example the number of pages in this book), it is possible to do this with complete accuracy.

acknowledge that variability in measured values is an inherent feature of all experimental work. What we need to be able to do is identify and quantify the variation, otherwise the reliability of our experiment is likely to be questioned, and any conclusions drawn from the experiment may be of limited value. If it is possible to identify the main cause(s) of the variation in the experimental data, then we may be able to redesign the experiment in order to reduce the variability.

The uncertainty quantifies in a clear manner the amount of variation that has been found in a measured value. An alternative term to that of uncertainty is to speak of an experimental *error*. In the context of data analysis, an error does not refer to a mistake (such as recording the wrong number) but is another word used to express the spread of measured values of a physical quantity. In this book we will consistently use the word *uncertainty* in preference to that of *error*, as there is less risk of the word being interpreted incorrectly, though you will find them both used in other books on data analysis.

EXERCISE A

1. Fill an electric kettle with a measured amount of cold water (say 500 cm^3). Switch the kettle on and measure the time it takes for the water to come to the boil. Empty the kettle and repeat the experiment. How variable are the boiling times? What factors do you think are responsible for the variation in times?
2. If your wristwatch has a stopwatch option, or if you can borrow a stopwatch, take a small coin and let it fall from a fixed height (say 3 m). Measure the time it takes to fall. Repeat the experiment ten times. What are the minimum and maximum times? What do you think are the most important factors affecting the timings?

4.2 Uncertainty in a single measurement

In any experiment it is easy to make a mistake when making or recording a measurement. If you have made only one measurement of a quantity, you need to be fairly confident that it is reliable, otherwise a spurious value may affect the success of your whole experiment. It is much better to make the measurement at least once more in order to be satisfied that the measured value is repeatable. There are circumstances in which it is difficult to make repeated measurements of a quantity — for example when it is continually changing with time. In these situations you may have to accept that a single measurement is the best that you can do. In the coming sections we will discuss what factors affect the value that you quote for the uncertainty in a single measurement.

4.2.1 Resolution uncertainty

No instrument exists that can measure a quantity to infinitely fine resolution. Take, for example, the measurement of length using a metre rule. The finest markings on the scale of such a rule are usually separated by a distance of 1 mm. You could, if you were very careful, be able to measure a length to within 0.5 mm, but it would be unlikely that you would be able to do much better than this. If you required a 'better' measurement of length, you could use vernier callipers or a micrometer which (usually) are capable of resolutions of 0.1 mm and 0.01 mm respectively. There are other length-measuring instruments, based on devices such as the laser, that offer even better resolution. Nonetheless, all measurements are limited by the instrument you are using. If the quantity you are measuring is stable or at least varies slowly with time, it is reasonable to quote the uncertainty as one half the smallest division on the scale. For example, you could quote a length measured with a metre rule as (361.0 ± 0.5) mm. The $\pm$ sign is used as shorthand to indicate that the length lies somewhere in the range 360.5 mm to 361.5 mm. In the case of a thermometer, the smallest division of which is 2°C, a temperature could be (61 ± 1)°C. In general, the resolution limit of an instrument represents the smallest uncertainty that can be quoted in a single measurement of a quantity.

4.2.2 Reading uncertainty

While making a measurement it is possible that the quantity under investigation varies by considerably more than half the smallest division on the instrument. Imagine heating a beaker containing water. We may be using a thermometer that has a resolution of ± 1°C or even better. However, we notice that as the water is stirred (in an effort to ensure a uniform temperature throughout), the thermometer indicates a wide temperature variation. At one instant the thermometer indicates 36°C, the next, 33°C, then quickly it changes to 35°C. Quoting an uncertainty of ± 1°C in these circumstances would be to *underestimate* the experimental uncertainty. In this situation we must use some judgement to

decide a reasonable value for the uncertainty. In a o
ment of this kind, there are no 'hard and fast' rules
uncertainties, and we have to rely on our commo
estimate the reading uncertainty to be less than ± 5
than ± 1°C, then we should choose a compromise between these
two values.

4.2.3 Calibration uncertainty

The instruments that you use in a laboratory should have been calibrated at some time against a standard. Expensive pieces of measuring equipment often come with a calibration certificate
indicating how closely the instrument conforms to the standard. If
scientists around the world are trying to compare their measurements, they need to be sure that their instruments 'agree' on what
is a metre, a volt, a second or whatever. The calibration certificate
provided by a manufacturer or a 'bureau of standards' is unlikely to
guarantee that the calibration is valid for more than 1 year. For the
calibration to remain valid, the instrument must be checked regu-
larly.

It is unlikely that you will have the time or facilities to make a
thorough check of the calibration of instruments you are using in
an experiment. However, a quick comparison of, say, voltages indicated by two voltmeters measuring the same voltage, or two thermometers which should be indicating the same temperature, may
save you from having to repeat a whole experiment due to an
uncalibrated, or perhaps faulty, piece of measuring equipment. An
uncalibrated, or a poorly calibrated instrument, leads to *systematic*
uncertainty in data and influences all measurements made with
that instrument. We will have more to say on systematic uncertainties in section 4.4.

4.2.4 Summary

It is important to be aware of resolution, reading and calibration
uncertainties when attempting to quote the uncertainty in a single
measurement. Such uncertainties exist whenever a measurement
is made in an experiment. However, to be able to get a real *feel* for
the variability in measurement, more than one measurement
should be made of each quantity. Where this is possible we can use
some results from statistical analysis to allow us to quantify experimental uncertainties. Let us look at probably the most important
quantity that can be calculated from a series of repeated measurements: the *mean*.

4.3 The mean

Look again at the data given in section 4.1.1 for the time of fall of an object:

Time (s)	0.64	0.61	0.63	0.53	0.59	0.65	0.60	0.61	0.64	0.71

We could expect the time that it really took for the object to fall to lie somewhere between the two extreme measured values, namely between 0.53 s and 0.71 s. If a single value for the time of fall is required, we can do no better than to calculate the average of the ten measurements that were made. The average of a group of numbers is commonly termed the *mean* of the numbers. The symbol used for the mean is $\bar{x}$ and it is calculated using the formula:

$$\bar{x} = \frac{\sum_{i=1}^{i=n} x_i}{n} \tag{4.1}$$

where n is the number of measurements made.

Σ is a *summation* sign which instructs us to add all of the x's together, so that:[2]

$$\sum_{i=1}^{i=n} x_i = x_1 + x_2 + x_3 + \dots\dots + x_n$$

Using the experimental data given in the table above, the mean time is given by:

$$\bar{x} = \frac{(0.64 + 0.61 + 0.63 + 0.53 + 0.59 + 0.65 + 0.60 + 0.61 + 0.64 + 0.71)\ s}{10}$$

$$= \frac{6.21\ s}{10} = 0.621\ s$$

We could quote the mean to one, two or three significant figures, that is 0.6 s, 0.62 s or 0.621 s. Which do we choose? This question is difficult to answer unless we have an estimate for the uncertainty in the mean value. We will not the use the methods we introduced in section 4.2 to estimate the uncertainty in a single measurement. Instead we will use the variability in the gathered data as a guide to estimating the uncertainty in the data.

---

2. x_1 refers to the first measured value, x_2 the second measured value and so on.

EXERCISE B

1. In a fluid flow experiment the amount of v
   through a pipe in a fixed time is measured. Giv
   measured volumes of water passing through the pipe, collected
   over eight successive 2-minute periods.

Volume (cm³)	48	45	45	50	47	46	43	44

   Calculate the mean volume collected in cm³, and the mean
   rate of flow in m³ s⁻¹.

2. The pressure of a vacuum system connected to a scanning electron microscope was monitored for a period of 2 hours and the
   following pressures recorded:

Pressure	3.2	7.5	8.0	6.5	1.0	0.9	1.3
(×10⁻³ Pa)	4.4	5.0	4.0	5.5	2.3	3.4	5.4

   Calculate the mean pressure over the 2-hour period.

4.3.1 Uncertainty in the mean

We will now consider a simple method by which the uncertainty
in the mean of a set of data can be calculated. In chapter 5 we will
deal with the matter of estimating uncertainties from gathered data
in more detail and discuss another method by which the uncertainty in the mean can be calculated.

In situations where, say, between five and ten repeat measurements have been made of a particular quantity, an estimate of the
uncertainty in the mean can be made by first calculating the *range*
of the data, where the range is given by:[3]

$$\text{range} = \text{largest value} - \text{smallest value}$$

The uncertainty in the mean may be found by dividing the range
by the number of measurements that were made, n:

$$\text{uncertainty in mean} = \frac{\text{range}}{n} \qquad (4.2)$$

---

3. In section 5.3 we will consider another method of calculating the uncertainty in the mean
   which gives values consistent with those obtained using equation 4.2.

As an example of the application of this method for calculating uncertainty, consider an experiment in which eight repeated measurements were made of the speed of sound in air (at 20°C). The recorded data are shown in table 4.2.

Table 4.2: Experimental values of the speed of sound in air at 20°C

Speed of sound $(m \ s^{-1})$	341.5	342.4	342.2	345.5	341.1	338.5	340.3	342.7

The mean of these numbers is 341.775 m s^{-1} and the range 345.5 − 338.5 = 7 m s^{-1}. By using equation 4.2, we find that the uncertainty is 7/8 = 0.875 m s^{-1}.

EXERCISE C

1. Using the data in table 4.1, calculate the uncertainty in the mean using equation 4.2.
2. In an experiment to study the properties of convex lenses, the distance from a lens to an image was measured on six occasions (object distance held fixed). The following data were obtained:

Image distance (mm)	425	436	428	417	429	413

Calculate the mean value of the image distance and the uncertainty in that value using equation 4.2.

4.3.2 How to quote uncertainties

In the previous example we might be tempted to say that the speed of sound in air, based upon the measurements we have made, is 341.775 m s^{-1} with an uncertainty of 0.875 m s^{-1}. We must be careful when we are quoting uncertainties. Their function is to quantify the *probable* range in which the value of that quantity lies. There is no point, therefore, in quoting the uncertainty to more than one significant figure.[4] In the present example we would round 0.875 m s^{-1} up to 0.9 m s^{-1}. Such a rounding indicates that the mean itself should not be quoted, as we have done, to

---

4. If the first figure in the uncertainty is a '1', such as in 1.356, it is usual to give the uncertainty to two significant figures — for example, in this case it would be ± 1.4.

three decimal places. In this situation we round the mean to the same number of decimal places as the uncertainty. In this example we would round 341.775 m s^{-1} to 341.8 m s^{-1}.

We can say that, based on data from the experiment, the value for speed of sound (symbol, c) in air = 341.8 m s^{-1} with an uncertainty of 0.9 m s^{-1}. This is rather a long-winded way of quoting the value and its uncertainty. A better way is to write it as:

$$c = (341.8 \pm 0.9) \text{ m s}^{-1}$$

The above method of quoting the value is quite acceptable. However, you will very often see the final value quoted in *scientific notation* (as discussed in section 2.5.3) — 341.8 becomes 3.418×10^2 and 0.9 becomes 0.009×10^2. Putting the two numbers together we get:

$$c = (3.418 \pm 0.009) \times 10^2 \text{ m s}^{-1}$$

To summarise, after making repeated measurements of a quantity, there are four important steps to take in quoting the value of the quantity:

Step 1
Calculate the mean of the measured values.

Step 2
Calculate the uncertainty in the quantity, making clear the method used. Round the uncertainty to one significant figure (or two if the first figure is a '1').

Step 3
Quote the mean and uncertainty to the appropriate number of figures.

Step 4
State the units of the quantity.

The uncertainty that we quoted in our example was 0.9 m s^{-1} which might suggest that all of our measurements should lie between (341.8 − 0.9) m s^{-1} and (341.8 + 0.9) m s^{-1}, that is, between 340.9 m s^{-1} and 342.7 m s^{-1}. Looking at the original data we see that this is *not* the case. Of the eight measurements shown in table 4.2, only *five* lie between 340.9 m s^{-1} and 342.7 m s^{-1}. Here we come to an important point.

When an uncertainty in an experimental value is quoted, we are not saying that the actual or *true* value of the quantity *must* lie between the limits given by (mean + uncertainty) to (mean −

uncertainty). The probability is high that it *will* lie between these limits, and we will see in chapter 5 that it is possible to quote a value for that probability.

EXERCISE D

Following are examples, drawn from a variety of student experiments, of quantities and their associated uncertainties. There is a mistake or omission in each of the statements. State the mistake that has been made and, where possible, correct it.

 (i) The spring constant = $(12.5731 \pm 0.628\ 41)$ N m^{-1}
 (ii) The density of copper = (8825 ± 500) kg m^{-3}
(iii) The velocity of the sound wave in water = $(1.48 \pm 0.05) \times 10^3$
(iv) The unknown capacitance = (1.1 ± 0.001) µF
 (v) The wavelength of the light is 6.0×10^{-7} m ± 1

4.3.3 Fractional and percentage uncertainties

In the majority of situations it is advisable to express the uncertainty in the same units as the quantity being measured. This is sometimes referred to as the *absolute* uncertainty in the quantity. In some cases, however, you may be required to state the ratio, $\dfrac{\text{uncertainty in quantity}}{\text{quantity}}$. This ratio is referred to as the *fractional* uncertainty in the quantity. For example, if the speed of an aircraft is (195 ± 5) m s^{-1}, then the fractional uncertainty in the speed is:

$$\frac{5 \text{ m s}^{-1}}{195 \text{ m s}^{-1}} = 0.026$$

It is important to note that, as the fractional uncertainty is a ratio of two quantities that have the same units, the fractional uncertainty itself has *no* units.

The *percentage* uncertainty expresses the uncertainty as a percentage of the original quantity. It is found by multiplying the fractional uncertainty by 100%. So, in the above example, the percentage uncertainty in the speed of the aircraft is:

$$0.026 \times 100\% = 2.6\%$$

As with absolute uncertainty, it is seldom necessary to quote fractional or percentage uncertainties to more than one significant figure. In the example above we can reasonably round the percentage uncertainty to 3%.

EXERCISE E

Determine the following absolute uncertainties percentage uncertainties:

(i) $I = (2.0 \pm 0.2)$ A

(iii) $t = (1.2 \pm 0.$

(ii) $r = (1.56 \pm 0.07)$ m

(iv) $m = (5.6 \pm 0.6) \times 10^2$ kg

4.3.4 True value, accuracy and precision

In everyday language, accuracy and precision signify much the same thing. In science and engineering, however, they have come to have different meanings. The difference can be explained if we first acknowledge that when we make a measurement of a quantity we are attempting to find an estimate of the 'true' value of that quantity. How many measurements do we need to make before we find the exact, or *true* value of a quantity? The answer is that the true value can never be known with absolute precision, but by gathering more data, and then finding the mean of the data values, we hope to get a better and better *estimate* of the true value. If our estimate, based on taking the mean of our data, is close to the true value, then we say that the measurements are *accurate*.

Although the true value can never be known, it should be possible for workers in various laboratories endeavouring to measure the same quantity (for example the thermal conductivity of a new alloy) to agree that the true value lies within certain limits.[5] By better experimental method, improved instruments or repeating the measurements many times, it is possible to set about narrowing the uncertainty limits. 'Best' values and uncertainties obtained for a quantity may be compared with those of other scientists and engineers with a view to establishing consistency between workers. If consistency is not found, the methods and materials used by all the workers need to be scrutinised closely. Some quantities are of sufficient importance that intensive study has reduced the uncertainty in the true value to a remarkable extent. For example, the electrical charge carried by an electron is known to be $(-1.602\ 177\ 3 \pm 0.000\ 000\ 5) \times 10^{-19}$ C (that is an uncertainty, expressed as a percentage, of 0.000 03%!).

When we speak of a measurement being *precise* we mean that the uncertainty in the value is small but this does not imply that it

---

5. We will see in chapter 5 that we can quote a probability that the true value of a quantity lies between certain limits.

is close to the true value. This may seem odd, and even contradictory. How *can* a measurement be precise but not give a value close to the true value? Let us consider a particular example:

Thermocouples are widely used to measure temperature and may be connected to a modified digital voltmeter so that a direct reading of temperature can be made. In an experiment, high purity water was poured into an open container and a heating element was used to bring the water to boiling point. Once the water was boiling, ten measurements of the temperature of the water were made using a thermocouple, and the measured values are shown in table 4.3.

Table 4.3: Ten repeat measurements of the boiling point of water, made using a thermocouple

Temp. (°C)	92.4	92.6	92.6	92.3	92.4	92.7	92.4	92.4	92.5	92.6

Using the method we discussed in section 4.3.1, we first calculate the mean of the above values and find it to be 92.49°C. The range of values is 0.4°C, so that the uncertainty is 0.4°C/10 = 0.04°C. Therefore, we quote the measured boiling point of water as:

$$\text{boiling point of water} = (92.49 \pm 0.04)°C$$

This looks fine except for one thing: the boiling point of pure water (at sea level, where the measurements were made) should be close to 100°C and all of our data are lower than this by more than 7°C. Our value for the boiling point has good precision, which implies that all the measured values are clustered tightly around the mean. However, it appears that the mean is some way away from the true value. Our experiment is precise, but it is *not* accurate. The highest priority must be given to establishing an accurate measure of the quantity under investigation, because precision without accuracy can only be misleading.

In a repeat of the previous experiment, a simple mercury-in-glass thermometer replaced the thermocouple. The scale on the thermometer could be read to the nearest 0.5°C. The data obtained are shown in table 4.4.

Table 4.4: Ten values of the boiling point of water, obtained using a mercury-in-glass thermometer

Temp. (°C)	101.0	100.5	99.0	99.0	99.5	100.5	100.0	101.0	100.5	101.0

The mean of this set of data is 100.2°C and the range is 2°C. It follows that the uncertainty in the boiling point of the water,

measured using the mercury thermometer, is 2°
write:

$$\text{boiling point of water} = (100.2 \pm 0.$$

The measurement of the boiling point of the v
precise using the mercury thermometer compared to the thermocouple (an uncertainty in the mean of 0.2°C compared to 0.04°C).
In terms of accuracy, however, the mercury thermometer has
proved (in this experiment) to be superior to that of the thermocouple. So why were the measured values for the temperature of
the boiling water found using the thermocouple so much lower
than 100°C? In answering that we must introduce the idea of systematic uncertainties. Before we deal with that topic, table 4.5
summarises what we have learned about the difference between
precision and accuracy.

Table 4.5: Summary of difference between accuracy and precision

If a value obtained from an experiment is:	...Then it is:
accurate	close to the true value but, unless given, the uncertainty could be of any magnitude
precise	it has a small uncertainty, but this does not mean that it is close to the true value
both accurate **and** precise	close to true value and with a small uncertainty. We would like our experimental data to fall into this category

4.4 Systematic and random uncertainties

Why did the values obtained using the thermocouple in section
4.3.4 differ so greatly from the expected value of 100°C? To answer
this question we must look more closely at the types of uncertainties that can occur in an experiment. Broadly speaking, we can
place uncertainties into two categories, referred to as *systematic*
and *random* uncertainties. We will come to random uncertainties
shortly, but it is worth stating now that random uncertainties are
usually easier to deal with than systematic uncertainties because
they are more easily identifiable and can be quantified using some
basic statistical reasoning.

Systematic uncertainties, on the other hand, tend to be more
serious as they are often difficult to detect and it is possible to

perform an experiment while being unaware of their existence. In our example in section 4.3.4, in which we were attempting to measure the boiling point of water, we were suspicious of the measurements — not because the measured values showed wide variation, but because they were so far away from the expected value of 100°C. Once systematic uncertainties have been detected they can be dealt with, but it is the detection that provides the challenge.

Two types of systematic uncertainty which can exist with measuring instruments are the *offset* uncertainty and *gain* uncertainty.

4.4.1 Offset uncertainty

Consider an experiment performed to find the melting point of water using the thermocouple mentioned in section 4.3.4 (which indicated a temperature of ~92.5°C when placed in boiling water). The same thermocouple was placed in a pure ice and pure water mixture and ten measurements of temperature were made. Table 4.6 shows the data obtained.

Table 4.6: Ten values of the melting point of water obtained using a thermocouple

Temp. (°C)	−7.5	−7.3	−6.9	−7.4	−7.4	−7.7	−7.6	−7.6	−7.3	−7.6

The mean of the above numbers is −7.43°C and the range is 0.8°C. Using our method for calculating the uncertainty in the mean, that is, uncertainty $= \dfrac{\text{range}}{n}$, we find that the melting point of water is (-7.43 ± 0.08)°C. Clearly there is a problem here: the melting point of pure water should be very close to 0.0°C. What we have uncovered is an offset uncertainty with our temperature-measuring system of about 7.5°C. For whatever reason (low battery, malfunctioning digital meter, incorrect type of thermocouple), all measurements of temperature are too low by about 7.5°C. An offset of this size may not be of much importance in the situation where you are measuring the temperature of a furnace which has been set to around 1500°C. By contrast, if you were trying to establish the body temperature of a new-born baby, a systematic error of 7.5°C would most certainly *not* be acceptable.

4.4.2 Gain uncertainty

Another cause of systematic uncertainty can be attributed to gain uncertainty in the measurement system. In contrast to the offset uncertainty, which remains fixed irrespective of the magnitude of the quantity being measured, the gain *is* dependent on the magnitude of the quantity. The effect of a gain uncertainty is best illustrated by example.

Five calibration masses (the values of which are known to high precision and accuracy) were placed in turn upon a top-loading electronic balance and the mass indicated by the balance was recorded. Table 4.7 shows the recorded values:

Table 4.7: Comparison between calibration masses and mass indicated by an electronic balance

Calibration mass (g)	0.00	20.00	40.00	60.00	80.00	100.00
Value indicated on balance (g)	0.00	20.26	40.65	60.98	81.20	101.52

We can see that as the mass placed on the balance increases, so the *difference* between measured and calibrated mass increases. If we use the symbol m_c to represent the calibrated mass and the symbol m_m to represent the measured mass, then table 4.8 shows the calibration mass and the difference between m_c and m_m.

Table 4.8: Difference between measured mass and calibration mass

m_c (g)	0.00	20.00	40.00	60.00	80.00	100.00
$(m_m - m_c)$ (g)	0.00	0.26	0.65	0.98	1.20	1.52

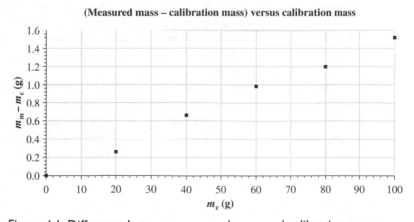

(Measured mass – calibration mass) versus calibration mass

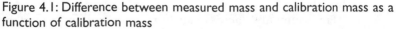

Figure 4.1: Difference between measured mass and calibration mass as a function of calibration mass

Figure 4.1 shows that the difference in the two quantities, $m_m - m_c$, increases in direct proportion to the magnitude of the mass located on the balance. As the relationship between the calibrated mass and the measured mass has now been established (at least over the range 0 to 100 g), future measurements of mass using this balance can be corrected for the gain uncertainty.

4.4.3 Detecting and dealing with systematic uncertainties due to instruments

When an instrument is used to measure a quantity, it is possible for that instrument to introduce a systematic uncertainty into your measurements. To properly identify offset and systematic gain uncertainties, we must check the values indicated by the instrument against 'known standards' across the range over which we wish to make measurements. If all the measured values differ from the standard values by the same amount, we have identified an offset uncertainty. As we saw in section 4.4.2, a gain uncertainty depends on the magnitude of the quantity being measured. Once the offset or gain uncertainty has been quantified, the measured values can be adjusted accordingly.

In short: to eliminate systematic uncertainties introduced by a measuring instrument, we must *calibrate* it using standards that are known to high precision *and* accuracy.

4.4.4. Random uncertainties

Random uncertainties produce scatter in observed values. The cause of the scatter could be due to the limitation in the scale of the instrument so that, on some occasions, measured values are rounded up above the 'true' value, and at other times the values are rounded down below the true value.

In some circumstances the measurement is difficult to perform and this causes a large variation in the measured values. For example, in an optics experiment, it is frequently the case that it is difficult to find the position of 'best focus' of an image on a screen. The screen is moved back and forth in an effort to find the position where the image is sharpest. Although it may be possible to read the measuring scale to ± 0.5 mm, or better, the uncertainty in the image position (in some cases) could easily be many times this value.

Environmental factors can introduce random uncertainties into measured values. For example:
• Electrical interference caused by switching on and off electrical equipment can affect sensitive voltage or current measurements.

- Vibrations caused by a passing motor vehicle (or even a passing person!) can unpredictably influence force measurements carried out using a sensitive electronic balance.
- Power supply fluctuations can affect the intensity of light being emitted from a lamp, which in turn influences optical measurements made using the lamp.
- In a water-flow experiment, changes in mains water pressure will cause variations in water-flow rates.

Although it may not be possible to reduce the random scatter in data, we can use statistical techniques to give us an estimate of the probable uncertainty and to allow us to calculate the effect of combining uncertainties. We will discuss the statistical approach to the addition of uncertainties in chapter 5. There are many situations in which we have insufficient data to justify a full statistical analysis, but still need to be able to estimate the effect of combining uncertainties. We will now consider methods by which uncertainties can be combined.

4.5 Combining uncertainties

An experiment may require the determination of several quantities which are later to be inserted into an equation. For example, we may measure the mass, m, of a body and its volume, V. The density of the body, ρ, can be calculated using the relationship:

$$\rho = \frac{m}{V} \tag{4.3}$$

How do uncertainties in the measured quantities, m and V, combine to give an uncertainty in ρ? The combination of uncertainties to give an uncertainty in a calculated value is sometimes called the *propagation* of uncertainties, or *error propagation*.

4.5.1 Symbols

Suppose the symbol x is used to represent a quantity (it could be a distance, for example). There are several symbols that are used to represent the uncertainty in x. These include:

$$\delta x, \Delta x, \varepsilon_x, \sigma_x \text{ and } \sigma_{\bar{x}}$$

In this chapter we will use Δx (pronounced *delta x*) to represent an uncertainty in x, although after discussion of the application of statistics to data analysis in chapter 5, we will see that there are good reasons for using the symbol $\sigma_{\bar{x}}$ (normally pronounced as *sigma-x bar*).

4.5.2 Combination of uncertainties: method I

This method is the most straightforward and requires only simple arithmetic. Each quantity in the formula is modified by an amount equal to the uncertainty in the quantity to produce the largest value and smallest value.

Example 1

Consider the calculation of the area of a wire of circular cross-section which has a measured diameter of (2.5 ± 0.1) mm. What is the area of cross-section and the uncertainty in that area?

The formula for the area, A, of a circle in terms of its diameter, d, is:

$$A = \frac{\pi d^2}{4} \tag{4.4}$$

Converting the diameter to metres and substituting into equation 4.4 gives:

$$A = \frac{\pi(2.5 \times 10^{-3} \text{ m})^2}{4}$$
$$= 4.91 \times 10^{-6} \text{ m}^2$$

To find the maximum cross-sectional area of the wire, we substitute the maximum value for the diameter, namely $(2.5 + 0.1)$ mm, that is 2.6 mm. The maximum area, A_{max}, is given by:

$$A_{max} = \frac{\pi(2.6 \times 10^{-3} \text{ m})^2}{4}$$
$$= 5.31 \times 10^{-6} \text{ m}^2$$

To find the minimum area A_{min}, we subtract the uncertainty from the quantity, so that:

$$A_{min} = \frac{\pi(2.4 \times 10^{-3} \text{ m})^2}{4}$$
$$= 4.52 \times 10^{-6} \text{ m}^2$$

The best estimate of the area is 4.91×10^{-6} m^2, with an upper value of 5.31×10^{-6} m^2 and lower value of 4.52×10^{-6} m^2.

We can quote the uncertainty succinctly by first subtracting A_{max} from A_{min} to give the *range* of A. This gives 0.79×10^{-6} m^2. Now divide the range by 2 to give 0.395×10^{-6} m^2. We round this to 0.4×10^{-6} m^2 and finally quote the area as:

$$A = (4.9 \pm 0.4) \times 10^{-6} \text{ m}^2$$

Example 2

In an electrical experiment, the current through a resistor was found to be (2.5 ± 0.1) mA and the voltage across the resistor (5.5 ± 0.3) V. Calculate the resistance of the resistor using $R = \dfrac{V}{I}$, and the uncertainty in R, ΔR.

$$\text{The resistance, } R = \frac{V}{I} = \frac{5.5 \text{ V}}{2.5 \times 10^{-3} \text{ A}}$$
$$= 2.20 \times 10^3 \ \Omega$$

When dealing with the quotient of two quantities,[6] as in this example, the maximum value of the quotient will occur when the numerator is increased by an amount equal to the uncertainty in the quantity appearing in the numerator, and the denominator is *reduced* by an amount equal to the uncertainty in the quantity in the denominator. To find the minimum value of the quotient we simply use the minimum value for the numerator and the maximum value for the denominator.

Here the maximum value of the numerator is $(5.5 + 0.3)$ V and the minimum value of the denominator is $(2.5 - 0.1) \times 10^{-3}$ A. Therefore, the maximum resistance R_{max}, is given by:

$$R_{max} = \frac{5.8 \text{ V}}{2.4 \times 10^{-3} \text{ A}} = 2.42 \times 10^3 \ \Omega$$

R_{min} is found by using the minimum value of voltage and the maximum value of current. This gives $R_{min} = 2.00 \times 10^3 \ \Omega$. To calculate the uncertainty in resistance, ΔR, we have:

$$\frac{R_{max} - R_{min}}{2} = 0.210 \times 10^3 \ \Omega$$

Rounding to the appropriate number of significant figures, we write the resistance as:

$$R = (2.2 \pm 0.2) \times 10^3 \ \Omega$$

EXERCISE F

1. The radial acceleration, a, of a body rotating in a circle of radius, r, at constant speed, v, is given by:

$$a = \frac{v^2}{r}$$

If $v = (3.00 \pm 0.05)$ m s^{-1} and $r = (1.5 \pm 0.1)$ m, calculate a, the maximum and minimum values of a, and the uncertainty in a.

---

6. The quotient of two numbers is simply the result of dividing one number by another.

2. The speed, v, of a transverse wave moving along a stretched string is given by:

$$v = \left(\frac{T}{\mu}\right)^{\frac{1}{2}}$$

where T is the tension in the string, and μ is the mass per unit length of the string. If $T = (25 \pm 2)$ N and $\mu = (1.2 \pm 0.1) \times 10^{-2}$ kg m^{-1}, calculate v and the maximum and minimum values of v.

4.5.3 Combination of uncertainties: method II

Although the method just treated is quite reasonable and general for finding combined uncertainties, it is quite cumbersome, especially when the formula contains more than one quantity with experimental uncertainty. We now introduce another technique for combining uncertainties, which represents an application of differential calculus. This may sound daunting and overly complicated but, in fact, if we can differentiate functions such as sines, cosines and logs, we will encounter few difficulties in calculating uncertainties involving these functions.

4.5.3.1 Partial differentiation

Suppose V depends on the two variables, a and b. The mathematical way of writing this is:

$$V = V(a, b)$$

We say the V is a *function* of a and b. An example of such a function would be:

$$V = ab^2$$

If a changes by an amount δa, and b changes by an amount δb, we can write the accompanying change in V, δV, as:

$$\delta V = \frac{\partial V}{\partial a}\delta a + \frac{\partial V}{\partial b}\delta b \tag{4.5}$$

$\frac{\partial V}{\partial a}$ is the partial derivative of V with respect to a. When finding a partial derivative, all quantities in the equation, except for the one that is being differentiated, are taken to be constant.

Examples

Consider $V = a^2b$. To find $\frac{\partial V}{\partial a}$, we treat b as a constant and a^2 differentiates to $2a$, to give $\frac{\partial V}{\partial a} = 2ab$.

To find $\dfrac{\partial V}{\partial b}$, we treat a as a constant and b differentiates to 1.

This gives $\dfrac{\partial V}{\partial b} = a^2$

In order to use equation 4.5 in problems of error propagation, we replace the quantities δa and δb by the *uncertainties* Δa and Δb, so that the equation is rewritten:

$$\Delta V = \left|\frac{\partial V}{\partial a}\right| \Delta a + \left|\frac{\partial V}{\partial b}\right| \Delta b \qquad (4.6)$$

$\left|\dfrac{\partial V}{\partial a}\right|$ means that we take the magnitude of the partial differential, that is, we ignore any minus sign that may occur once we have differentiated. The consequence of not doing this is that cancellation of the terms on the right-hand side of equation 4.6 could occur. Note that we partially differentiate the function with respect to each quantity that possesses uncertainty. Quantities with no uncertainty are regarded as constants.

EXERCISE G

(i) If $s = \frac{1}{2}at^2$, calculate $\dfrac{\partial s}{\partial a}$ and $\dfrac{\partial s}{\partial t}$

(ii) If $P = I^2R$, calculate $\dfrac{\partial P}{\partial I}$ and $\dfrac{\partial P}{\partial R}$

(iii) If $n = \dfrac{\sin i}{\sin r}$, calculate $\dfrac{\partial n}{\partial i}$ and $\dfrac{\partial n}{\partial r}$

(iv) If $v = \left(\dfrac{T}{\mu}\right)^{\frac{1}{2}}$, calculate $\dfrac{\partial v}{\partial T}$ and $\dfrac{\partial v}{\partial \mu}$

4.5.4 Combining uncertainties: sums, differences, products and quotients

Equation 4.6 is applicable to any formula that you are likely to encounter. However, there are situations that are so common, such as taking the product of two quantities, that it is worth using equation 4.6 to determine the relationship that combines the uncertainties in the quantities.

Sum

If $V = a + b$, and uncertainties in a and b are Δa, and Δb respectively, we can use equation 4.6 to find the uncertainty in V. We have:

$$\Delta V = \left|\frac{\partial V}{\partial a}\right|\Delta a + \left|\frac{\partial V}{\partial b}\right|\Delta b$$

Now $\left|\dfrac{\partial V}{\partial a}\right| = 1$ and $\left|\dfrac{\partial V}{\partial b}\right| = 1$, so that:

$$\boxed{\Delta V = \Delta a + \Delta b}$$

We see that the uncertainty in V is given by the sum of the uncertainties in a and b.

Difference

If $V = a - b$ then once again, $\left|\dfrac{\partial V}{\partial a}\right| = 1$ and $\left|\dfrac{\partial V}{\partial b}\right| = 1$ so that:

$$\boxed{\Delta V = \Delta a + \Delta b}$$

Thus, when dealing with a difference, the uncertainty in V is the *sum* of the uncertainties in a and b.

Product

If $V = ab$, then $\left|\dfrac{\partial V}{\partial a}\right| = b$ and $\left|\dfrac{\partial V}{\partial b}\right| = a$, so that using equation 4.6 we get, $\Delta V = b\Delta a + a\Delta b$. If we divide both sides of this equation by ab we get:

$$\frac{\Delta V}{ab} = \frac{b\Delta a}{ab} + \frac{a\Delta b}{ab}$$

so that:

$$\boxed{\frac{\Delta V}{V} = \frac{\Delta a}{a} + \frac{\Delta b}{b}}$$

The uncertainty in the quantity divided by the quantity itself, in this case $\dfrac{\Delta V}{V}$, is the *fractional* uncertainty in V and is equal to the fractional uncertainty in a added to the fractional uncertainty in b.

Quotient

If $V = \dfrac{a}{b}$, then $\left|\dfrac{\partial V}{\partial a}\right| = \dfrac{1}{b}$ and $\left|\dfrac{\partial V}{\partial b}\right| = \dfrac{a}{b^2}$ so that equation 4.6 becomes, $\Delta V = \dfrac{\Delta a}{b} + \dfrac{a\Delta b}{b^2}$. If both sides are divided by $\dfrac{a}{b}$, then we find that, as in the case of the fractional product ab:

$$\boxed{\frac{\Delta V}{V} = \frac{\Delta a}{a} + \frac{\Delta b}{b}}$$

Example 1

The temperature of $(3.0 \pm 0.2) \times 10^2$ g of water is raised by $(5.5 \pm 0.5)°C$ by a heating element placed in the water. Calculate the amount of heat transferred to the water to cause this temperature rise. Also calculate the *uncertainty* in the amount of heat transferred to the water.

The relevant formula that relates heat input, Q, to temperature rise, θ, is:

$$Q = mc\theta \tag{4.7}$$

where m is the mass of the water and c is its specific heat capacity. We can find a value for c by referring to a data book such as that by Kaye and Laby (full details in appendix 1). The value of c is 4186 J kg^{-1} °C^{-1}. Assuming that the uncertainty in this value is small enough to be ignored, we can write:

$$\Delta Q = \left|\frac{\partial Q}{\partial m}\right| \Delta m + \left|\frac{\partial Q}{\partial \theta}\right| \Delta \theta \tag{4.8}$$

Now $\left|\dfrac{\partial Q}{\partial m}\right| = c\theta$ and $\left|\dfrac{\partial Q}{\partial \theta}\right| = mc$, so that equation 4.8 becomes (converting the mass to kilograms):

$\Delta Q = c\theta\Delta m + mc\Delta\theta$
$= 4186$ J kg^{-1}°C$^{-1} \times 5.5°C \times 0.02$ kg $+ 0.3$ kg $\times 4186$ J kg^{-1} °C$^{-1} \times 0.5°C$
$= 460.5$ J $+ 627.9$ J $= 1088.4$ J

Now we use equation 4.7 to calculate the heat, Q:

$$Q = 0.3 \text{ kg} \times 4186 \text{ J kg}^{-1}°C^{-1} \times 5.5°C = 6907 \text{ J}$$

We can now quote the energy transferred as $Q = (6.9 \pm 1.1) \times 10^3$ J.

Example 2

In section 4.5.2 the uncertainty in the cross-sectional area, A, of a wire was calculated, given that the diameter of the wire was (2.5 ± 0.1) mm. The area was found to be $(4.9 \pm 0.4) \times 10^{-6}$ m^2. Use the method given above to calculate the uncertainty in A.

75

We have that:

$$A = \frac{\pi d^2}{4}$$

ΔA is the uncertainty in A and this is found by writing equation 4.6 as:

$$\Delta A = \left|\frac{\partial A}{\partial d}\right| \Delta d$$

Now

$$\left|\frac{\partial A}{\partial d}\right| = \frac{2\pi d}{4} = \frac{\pi d}{2}$$

so that,

$$\Delta A = \frac{\pi d \Delta d}{2} = \frac{\pi \times 2.5 \times 10^{-3}\ \text{m} \times 0.1 \times 10^{-3}\ \text{m}}{2}$$

$$= 3.9 \times 10^{-7}\ \text{m}^2$$

Rounding this to one significant figure gives 4×10^{-7} m², which is equal to the value found for the uncertainty in section 4.5.2.

EXERCISE H

In the following exercises remember to express your answer to the correct number of significant figures.

1. The magnetic field, B, at the centre of a solenoid with n turns per metre through which a current I passes is given by $B = \mu_0 nI$, where μ_0 is the permittivity of free space ($= 4\pi \times 10^{-7}$ H m⁻¹). If $n = (400 \pm 5)$ m⁻¹ and $I = (3.2 \pm 0.2)$ A, calculate B and the uncertainty in B, ΔB.

2. The current density, j, through a conductor of cross-sectional area, S, is given by, $j = \frac{I}{S}$. If $I = (2.6 \pm 0.2)$ A and $S = (2.5 \pm 0.1) \times 10^{-6}$ m², calculate j and the uncertainty in j, Δj.

3. The focal length of a lens, f, is related to object distance, u, and image distance, v, by, $\frac{1}{f} = \frac{1}{u} + \frac{1}{v}$. If $u = (30.0 \pm 0.1)$ cm and $v = (20.0 \pm 0.5)$ cm, calculate f and the uncertainty in f, Δf.

4. The number of active nuclei, N, that remain in a sample of radioactive material after a time, t, is given by, $N = N_0 \exp(-\lambda t)$, where N_0 is the number of active nuclei when $t = 0$, and λ is referred to as the decay constant. If $N_0 = (6.0 \pm 0.4) \times 10^6$ and $t = (1.28 \pm 0.02) \times 10^3$ s, calculate N and the uncertainty in N, ΔN, given that $\lambda = 5.2 \times 10^{-3}$ s⁻¹.

4.6 Selection and rejection of data

Data selection and rejection is a sensitive subject and one that can bring out strong feeling amongst experimenters. There are those who would argue that 'all data are equal' and that there are no circumstances in which the rejection of data can be condoned. At the other extreme there are those that 'know' that a set of data is spurious and reject it without a second thought. It is possible to feel sympathy for both views: if you have spent days setting up an experiment and have confidence in the technique you are using, why *would* you reject any of the data you have collected? On the other hand, if you have operated the equipment many times and are familiar with the sort of numbers to expect from an experiment, it is not surprising that a set of data differing considerably from the rest is labelled as 'spurious', and forgotten about.

There are statistical tests that can be applied to data which will assist in selecting data for rejection. However, the application of such tests, especially if done automatically by computer, can throw out data before anyone has had time to ask the question 'Are the data points *truly* spurious, or is there something going on that I should know about?'.

I consider that all data should be recorded as the experiment proceeds and that human (or computer) intervention that would have the effect of 'filtering' the data should be avoided. This is not to say that all the data gathered should be used in further analysis. For example, if your concentration lapsed while you were timing an event 'by hand' it would be quite legitimate to neglect that timing in later calculations.

Confidence in an experiment, and data obtained from that experiment, really comes when you are satisfied that the experiment is repeatable. If you *do* have a suspect data point, but you can see no reason to reject it, then good advice is to repeat the experiment. This may not always be possible. The experiment may consist of testing something to destruction. For example, in an experiment to study the relationship between stress applied to a wire, and the strain produced in that wire,[7] you may be required to stretch the wire until it breaks. You can take another 'identical' wire to the first and repeat the experiment, but of course, by choosing a new wire, a very important element of the experiment has changed.

---

7. Stress is proportional to the force applied to the ends of the wire. Strain is proportional to the change in length of the wire that occurs as a consequence of the application of stress.

4.7 Comment

Uncertainties in data are a part of life for experimenters in science and engineering. In this chapter we have discussed the types of uncertainty that can occur during an experiment and methods by which they may be combined. Although the approach we adopted when combining uncertainties is good enough for many situations, it is possible that we are not getting the most from our data. In the next chapter we will study the distribution that provides a good description of variability for the majority of data that we will gather: the normal distribution. This will lead us to another method of combining uncertainties.

PROBLEMS

1. Eight measurements of the resistance of a mercury sample were made (at 300 K), the values obtained given in table 4.9.

   Table 4.9: Eight repeat measurements of the resistance of a sample of mercury

Resistivity ($\times 10^{-2}$ Ω)	9.5	9.3	9.9	9.9	9.1	8.9	9.6	9.3

   (i) What is the range of the data in table 4.9?
   (ii) Calculate the mean resistance and use the range to estimate the uncertainty in the resistance.
   (iii) Quote the mean and the uncertainty in the mean to the appropriate number of significant figures.

2. The ideal gas equation relates the pressure, P, volume, V and temperature, T (in kelvins) for any gas at low pressure. The equation is:

$$PV = nRT$$

   where R is a constant with value 8.314 J K^{-1} mol^{-1} and n is the number of moles of gas present.

   If $P = (0.6 \pm 0.1) \times 10^5$ Pa, $V = (22 \pm 2) \times 10^{-3}$ m^3, and $T = (325 \pm 5)$ K, calculate n and the uncertainty in n (assume that the uncertainty in R is negligible).

3. As part of a study on the quality of river water, the following characteristics of the river water were measured: Flow rate, pH, temperature, electrical conductivity and lead content. Table 4.10 shows seven repeat measurements of each of the quantities.

Table 4.10: River water data

Flow rate (m s^{-1})	pH	Temperature (°C)	Electrical conductivity (μS cm^{-1})	Lead content (ppb)*
0.45	6.9	10.0	906	37
0.48	6.8	11.0	1105	54
0.46	7.3	11.0	1004	34
0.39	7.2	9.5	998	45
0.41	7.2	10.0	885	68
0.41	6.9	10.5	780	56
0.46	7.0	11.5	885	70

* ppb stands for 'parts per billion'.

(i) Calculate the mean and uncertainty in each of the quantities shown in table 4.10 using the method discussed in section 4.3.1.

(ii) Which quantity exhibits the largest *fractional* uncertainty?

4. The period of the motion, T, of a body of mass, M, on the end of a spring of spring constant, k, is given by:

$$T = 2\pi \sqrt{\frac{M}{k}}$$

Given that $M = (210 \pm 5)$ g and $T = (1.1 \pm 0.1)$ s:

(i) Calculate k.

(ii) Use the method described in section 4.5.2 to find the uncertainty in k.

(iii) Repeat the calculation of uncertainty using the 'partial differentiation method' described in section 4.5.3.1.

5. The gain, G, of an amplifier in decibels may be written:

$$G = 20 \log_{10} \left(\frac{V_{out}}{V_{in}} \right)$$

where V_{out} and V_{in} are the output and input voltages of the amplifier respectively. A student finds the output voltage of an amplifier to be (122 ± 3) mV and the input voltage to be (33 ± 2) mV. Calculate the gain of the amplifier and the uncertainty in the gain. (Hint: if you are going to use the method for combining uncertainties discussed in section 4.5.3.1, then it is useful to know that, if $y = \log_{10}(u)$, $\dfrac{\partial y}{\partial u} = \dfrac{0.4343}{u}$).

6. (i) Given that $z = a.e^b$, find $\dfrac{\partial z}{\partial a}$ and $\dfrac{\partial z}{\partial b}$.

   (ii) Using this information and given that $a = (4.6 \pm 0.2)$ and $b = (10.2 \pm 0.1)$, find z and the uncertainty in z.

7. Table 4.11 shows the times taken for a reaction to occur between sodium thiosulphate and hydrochloric acid (see problem 6 at the end of chapter 3). Included in the table are the uncertainties in the measurements of time.

Table 4.11: Time for reaction between $Na_2S_2O_3$ and hydrochloric acid as a function of concentration of $Na_2S_2O_3$

Concentration (M)	Reaction time, t (s)
0.20	25 ± 2
0.16	30 ± 2
0.14	32 ± 2
0.12	40 ± 3
0.08	56 ± 3
0.02	116 ± 3

There is a linear relationship between $\dfrac{1}{\text{reaction time}}$ and the concentration. Attach two columns to table 4.11, one showing $\dfrac{1}{\text{reaction time}}$ and the other, the uncertainty in this quantity.

80

STATISTICAL APPROACH TO VARIABILITY IN DATA

5.1 Overview: estimating uncertainties with the aid of statistics

We learned in chapter 4 that no matter how much care is taken during an experiment, or how sophisticated the equipment used, measured values are influenced by random uncertainties. It is useful to think of these uncertainties as concealing the 'true' value of the quantity being sought. Random uncertainties cause measured values to occur above and below the true value.

In this chapter we will consider methods 'borrowed' from statistics for dealing with variability in data. Statistics can be described as the science of assembling, organising and interpreting numerical data and so is an ideal tool for assisting in the analysis of experimental data.

In this chapter we will adopt the approach of basing our determination of the uncertainty in a quantity on the spread of data obtained through repeated measurements of a quantity, and not be concerned with assessing the uncertainties in individual measurements. This approach is valid when we have made sufficient measurements (say in excess of five) to satisfactorily describe the spread in data due to random uncertainties.

We will not attempt to derive the formulae given in this chapter but indicate through examples their credibility and applicability. Good books to refer to, which discuss the derivation of the formulae and deal extensively with the application of statistics to science and engineering, are those by Meyer and Taylor detailed in appendix 1.

5.2 Variance and standard deviation of repeated measurements

The mean of a set of data obtained by repeated measurements is regarded as the best estimate of the 'true' value of the quantity being measured, but how may the data be used to give a single number which represents the uncertainty in that best estimate? To answer that question we will begin by looking at a typical set of experimental data in which random uncertainties have caused a spread in the measured values.

In a study of the motion of a body as it slides down an inclined plane, the time for the body to slide 80 cm was measured using a hand held stopwatch. Table 5.1 shows values from ten repeat timings of the motion.

Table 5.1: Ten repeat measurements of time for a body to slide down a plane

Time (s)				
0.64	0.64	0.59	0.58	0.70
0.61	0.68	0.55	0.57	0.63

The mean of the numbers in table 5.1 is 0.619 s. To estimate the spread or variability in the data we begin by drawing up a table which indicates how far each measured value is from the mean. The difference between the ith data value and the mean is called the *deviation* and is represented by the symbol d_i. Table 5.2 shows the deviation and the square of the deviation from the mean of the data given in table 5.1.

Table 5.2: Measured values, deviations from mean and (deviations from mean)^2

x_i (s)	$d_i = x_i - \bar{x}$ (s)	$d_i^2 = (x_i - \bar{x})^2$ (s^2)
0.64	0.021	0.000 441
0.64	0.021	0.000 441
0.59	−0.029	0.000 841
0.58	−0.039	0.001 521
0.70	0.081	0.006 561
0.61	−0.009	0.000 081
0.68	0.061	0.003 721
0.55	−0.069	0.004 761
0.57	−0.049	0.002 401
0.63	0.011	0.000 121

$$\sum (x_i - \bar{x}) = 0 \qquad \sum (x_i - \bar{x})^2 = 0.020\,89 \text{ s}^2$$

At first sight it might seem reasonable to take the mean of the deviations shown in the second column of table 5.2 as a measure of the variability of the data, but there is a difficulty with this. As we can see from that column, the sum of the deviations is zero[1] so that the mean deviation is also zero. We need to look elsewhere for a number which represents the spread of the data.

What *is* most often taken as a measure of the variability of the data is the mean of the sum of the *squares* of the deviations. This is referred to as the *variance* of the data and is most usually represented by the symbol σ^2. The variance is defined as:

$$\sigma^2 = \frac{\sum d_i^2}{n}$$

so that, $\qquad\qquad\qquad \sigma^2 = \frac{\sum (x_i - \bar{x})^2}{n}$ (5.1)

where n is the number of repeat measurements.

The third column in table 5.2 shows the square of the deviations and their sum, which is $0.020\,89$ s^2. It follows that:

$$\sigma^2 = \frac{0.020\,89\ \text{s}^2}{10} = 0.002\,089\ \text{s}^2$$

Notice that the units of the variance are the *square* of the units in which the original measurements were made.

Although variance is a fundamental measure of the spread of data values, it is more common to find the *standard deviation* of a set of measurements presented as a measure of spread. This quantity has the same units as the original measurements. Standard deviation is defined as the (variance)$^{\frac{1}{2}}$ and is represented by the symbol σ. Using equation 5.1 we get:

$$\sigma = \left(\frac{\sum (x_i - \bar{x})^2}{n} \right)^{\frac{1}{2}}$$ (5.2)

For the data shown in table 5.1 the standard deviation is $(0.002\,089\ \text{s}^2)^{\frac{1}{2}} = 0.045\,71$ s.

It is worth noting that many pocket calculators have in-built statistical functions such as that given by equation 5.2. The use of these functions can aid calculations considerably, especially when there are many numbers to process. Chapter 8 discusses the use of pocket calculators for data analysis in some detail.

---

1. In fact, this is just another way of defining the mean.

If we were to repeat the timing experiment in section 5.2, but on this occasion make 50 repeat measurements, how would the mean and standard deviation be affected? Table 5.3 shows fifty consecutive measurements of time.

Table 5.3: Fifty consecutive measurements of time for a body to slide down a plane

Time (s)									
0.61	0.58	0.60	0.60	0.53	0.64	0.64	0.59	0.58	0.70
0.60	0.51	0.69	0.66	0.56	0.61	0.68	0.55	0.57	0.63
0.61	0.64	0.60	0.61	0.58	0.62	0.64	0.53	0.59	0.63
0.61	0.63	0.65	0.64	0.58	0.61	0.68	0.62	0.61	0.60
0.68	0.56	0.59	0.53	0.55	0.59	0.62	0.54	0.59	0.55

It is left as an exercise to show that the mean of the above values is 0.6042 s and that the standard deviation, calculated using equation 5.2, is 0.043 64 s.

Table 5.4 shows a comparison of the means and standard deviations for 10 and 50 measurements of the time for a body to slide down an inclined plane.

Table 5.4: Means and standard deviations compared for 10 and 50 measurements of the same quantity

	Mean, $\bar{x}$ (s)	Standard deviation, σ (s)
Ten measurements	0.6190	0.045 71
Fifty measurements	0.6042	0.043 64

Table 5.4 shows that the standard deviation is virtually unchanged by taking 50 measurements rather than 10. In general the following statement holds for the standard deviation: *The standard deviation of a set of repeat measurements of a quantity remains almost constant, regardless of how many measurements are made.*

A question arises: should we take the standard deviation as the uncertainty in the mean and, if so, what is the point in making a large number of repeat measurements if the standard deviation can be calculated using only a few?

Let us consider again the reason for making repeated measurements of a quantity during an experiment. We are trying to find the best estimate for the quantity which we take to be the mean but,

in addition, we would like an estimate of the
mean. The standard deviation is a number wh
of the spread of the *whole* data set and ther
taken as the uncertainty in the mean.

We introduce a new term, the *standard err*
resented by the symbol, $\sigma_{\bar{x}}$), when speaking
the mean. By reference to a specific example we will describe
what the standard error is and how it is related to the standard
deviation of repeated measurements.

5.3 Uncertainty in the mean of repeated measurements: standard error of the mean

In a fluid-flow experiment, 10 measurements were made of the
volume of water flowing through the apparatus in one
minute. Table 5.5 shows the data gathered.

**Table 5.5: Values from 10 repeat measurements of volume of water
collected in a fluid-flow experiment**

Volume of liquid collected (mL)									
33	45	43	42	45	42	41	44	40	42

The mean of the values in table 5.5 is 41.7 mL, with standard
deviation, calculated using equation 5.2, of 3.29 mL.

The experiment was performed eight times in all, with each
experiment consisting of 10 repeat measurements of volume. Table
5.6 shows the mean and standard deviation for each of the 8 sets
of measurements.

**Table 5.6: Means and standard deviations for 8 sets of measurements
(each set consists of 10 repeat measurements)**

Mean (mL)	41.0	41.7	40.4	41.5	41.7	40.4	42.5	39.5
Standard deviation (mL)	3.13	3.29	3.07	3.11	3.20	2.94	2.73	3.20

By comparing the means in table 5.6 with the 10 values in table
5.5 we see that the variability in the means is considerably less
than the variability in the original data. We should expect this

because the reason for calculating the mean is to 'average out' the scatter due to the random uncertainties which influence individual measurements.

We can take the values in table 5.6 and calculate the mean of the means, which we will represent by the symbol, $\bar{X}$, and the standard deviation of the means (calculated using equation 5.2), which is the standard error in the mean, $\sigma_{\bar{x}}$. The calculations give:

$$\bar{X} = 41.1 \text{ mL}$$

and
$$\sigma_{\bar{x}} = 0.893 \text{ mL}$$

$\bar{X}$ differs little from the individual means appearing in table 5.6. However, the standard error of the mean is less than the standard deviation in the original data by a factor of approximately 3. As the number of repeat measurements in each experiment is equal to 10, we can see that, for this example at least, we could write the relationship between $\sigma_{\bar{x}}$ and σ as:

$$\sigma_{\bar{x}} \approx \frac{\sigma}{\sqrt{n}} \tag{5.3}$$

where n is the number of repeat measurements[2].

A formal mathematical proof[3] would show that the $\approx$ symbol in equation 5.3 can be replaced by the = symbol. We can write:

$$\sigma_{\bar{x}} = \frac{\sigma}{\sqrt{n}} \tag{5.4}$$

It is $\sigma_{\bar{x}}$ that we take as the uncertainty in the mean.

We can now return to the data in table 5.4 and use equation 5.4 to obtain the uncertainty in the mean when 10 and 50 measurements are made.

The uncertainty in the mean of the 10 measurements of time is:

$$\sigma_{\bar{x}} = \frac{\sigma}{\sqrt{n}} = \frac{0.045\,71 \text{ s}}{\sqrt{10}} = 0.014 \text{ s}$$

---

2. For example, using the first standard deviation appearing in table 5.6 we get, $\frac{\sigma}{\sqrt{n}} = \frac{3.13 \text{ mL}}{\sqrt{10}} = 0.990$ mL which compares with the value of 0.894 mL found for the standard deviation of the means.

3. We will not attempt to derive this formula. See chapter 5 of *An Introduction to Error Analysis* by J. R. Taylor (full details in appendix 1) for a derivation.

The best estimate of the true time based on ten repeat measurements is:[4]

$$(0.619 \pm 0.014) \text{ s}$$

For the set of 50 measured values we have:

$$\sigma_{\bar{x}} = \frac{\sigma}{\sqrt{n}} = \frac{0.043\ 64 \text{ s}}{\sqrt{50}} = 0.0062 \text{ s}$$

The best estimate of the true value of the quantity measured is therefore:

$$(0.604 \pm 0.006) \text{ s}$$

We see that it *is* worthwhile to make many repeat measurements of a quantity if we want to reduce the uncertainty in the estimate of the true value of that quantity. However, to reduce the uncertainty by a factor of 2, equation 5.4 indicates that we must increase the number of measurements by a factor of 4.

To summarise:

(i) The best estimate of a quantity is found by taking the mean of a set of repeated measurements of that quantity.

(ii) The standard deviation is a measure of the spread of the measurements and is insensitive to how many measurements are made.

(iii) The standard error in the mean is taken as the uncertainty in the mean and this *does* decrease as the number of repeat measurements increases.

EXERCISE A

1. Consider the following set of numbers drawn from an experiment.

   25.2   23.3   22.7   27.0   26.8   24.0   26.6   22.9

   Calculate the mean, standard deviation and standard error in the mean of the above numbers. Give the mean and the uncertainty in the mean to the appropriate number of significant figures.

---

4. Although this is an alternative method for determining uncertainties to that introduced in chapter 4, the recommendation that uncertainties be quoted to one significant figure (unless the first figure is a '1') still applies.

2. A steel ball is dropped from a height of 400 mm onto a metal plate on 60 occasions. The rebound height of the ball is shown in table 5.7.

Table 5.7: Values obtained from 60 repeat measurements of the rebound height of a ball

Rebound height (mm)					
176	173	174	176	175	173
176	176	176	178	176	171
178	176	174	176	173	174
177	176	177	176	175	176
175	176	178	176	172	177
179	169	176	178	173	177
178	178	176	176	176	175
178	170	178	176	177	178
178	179	176	180	178	178
177	173	177	176	180	173

Using the numbers in table 5.7, calculate the mean, variance, standard deviation and standard error in the mean rebound height. Give the mean and the uncertainty in the mean to the appropriate number of significant figures.

5.4 Displaying the values obtained from repeated measurements: the histogram

A useful way of displaying data from many repeat measurements of a quantity is to draw a bar chart, often referred to as a *histogram*. The range of the data[5] is divided into a number of equal intervals and the number of values that fall into each interval[6] (referred to as the *frequency*) is plotted vertically with the intervals plotted horizontally.

Table 5.8 shows 90 values of the temperature of a water bath displayed in the form of a histogram.

When deciding how many intervals should appear in a histogram there are no hard and fast rules to be followed. A simple method of constructing a histogram is to make the number of

---

5. The range of the data is equal to the (maximum data value – minimum data value).
6. An interval in a histogram is sometimes referred to as a 'bin'.

intervals approximately equal to the square root of the number of values in the data set. The width of the intervals appearing along the horizontal axis are then made equal to the range divided by the number of intervals. In some cases (as in the example below) the width of the interval is rounded in order to make the counting of the number of value in each interval easier (even if this means that the total number of intervals increases or decreases slightly).

Table 5.8: Ninety repeat measurements of the temperature of a water bath (unit °C)

52.9	44.8	51.6	43.0	45.0	46.5	37.2	48.9	49.4
52.5	45.9	49.0	44.7	38.5	50.5	50.8	49.8	38.1
44.0	48.7	52.3	44.2	47.2	42.9	38.8	41.1	50.3
42.8	43.3	47.1	49.1	45.4	47.2	45.9	42.0	50.7
48.1	48.3	52.9	46.0	47.4	48.6	46.5	44.2	36.5
50.7	43.7	42.8	42.9	35.6	46.9	51.3	46.1	47.5
47.8	47.8	52.4	39.7	35.6	46.8	38.3	50.5	42.6
55.4	45.4	35.9	47.4	40.0	49.9	37.2	42.5	40.1
46.9	45.7	52.1	40.5	41.1	50.0	39.3	42.1	48.4
40.7	41.1	44.4	49.6	47.9	41.0	46.4	46.5	39.0

The first column in table 5.9 indicates the steps by which a histogram can be constructed. The second column shows those steps applied to the data in table 5.8.

Table 5.9: Step-by-step approach to plotting a histogram

Steps to take when plotting a histogram	Worked example based on data shown in table 5.8
1. Find the range, R, of the data (max. value – min. value)	$R = 55.4°C – 35.6°C = 19.8°C$
2. Count total number of values in data set, N	$N = 90$
3. Calculate $\sqrt{N}$ and round to the nearest whole number. This is taken as the number of intervals	$\sqrt{N} = 9.487$, rounds to 9
4. Divide R by number calculated in step 3 to give width of each interval	width of interval = 2.2, rounds to 2

(continued)

Table 5.9 (continued)

Steps to take when plotting a histogram	Worked example based on data shown in table 5.8	
5. Draw up table showing all the intervals covering the range and the number of data values (i.e. the frequency) in each interval. Make it clear into which interval a number on the borderline between two intervals (e.g. 45 in the worked example) should be placed. Note an effect of rounding the interval width to 2 is to increase the total number of intervals required from 9 to 11.	interval (°C)	frequency
	$35 \leq x < 37$	4
	$37 \leq x < 39$	6
	$39 \leq x < 41$	8
	$41 \leq x < 43$	11
	$43 \leq x < 45$	9
	$45 \leq x < 47$	15
	$47 \leq x < 49$	15
	$49 \leq x < 51$	13
	$51 \leq x < 53$	8
	$53 \leq x < 55$	0
	$55 \leq x < 57$	1

Figure 5.1 shows the data from table 5.8 presented in the form of a histogram.

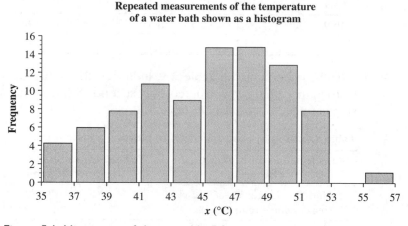

Figure 5.1: Histogram of data in table 5.8

The method of constructing a histogram is now applied to the timing data shown in table 5.3. The histogram is shown in figure 5.2.

The histogram in figure 5.2 displays the following features:

(i) The interval 0.58 s to 0.62 s has the largest frequency. This interval includes the mean and indicates that a large proportion of the measured values lie close to the mean.

(ii) The spread or *distribution* of values is approximately symmetrical about the interval containing the mean.

(iii) There are few values that lie far from the mean.

Characteristics (i) to (iii) are displayed by data gathered in such a wide variety of experiments that a brief discussion of a statistical distribution that possesses these properties is in order.

One distribution of data which exhibits the above characteristics is referred to as the *normal* distribution[7] and data from experiments are often referred to as being 'normally distributed'. The normal distribution can be derived from 'first principles' but we will not do that here. What we need to know are the features of the distribution which are useful when describing and analysing experimental data.

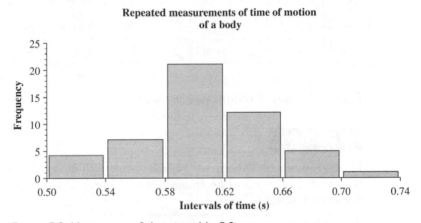

Figure 5.2: Histogram of data in table 5.3

5.4.1 Properties of the normal distribution

A large data set which is normally distributed would produce a histogram like that shown in figure 5.3. The intervals along the horizontal axis have been chosen to be very narrow and, to give a better idea of the shape of the distribution, a line is drawn which passes through the top of each bar on the histogram, all other lines being omitted. We can see that the line has a 'bell' shape with a peak at the mean, $\bar{x}$. These are features common to all data which follow the normal distribution.

The standard deviation, σ, is taken to be the characteristic width of the normal distribution. Figure 5.3 shows the central region of

---

7. Also referred to as the Gaussian distribution.

the normal curve bounded by two vertical lines, one drawn at the x-value $\bar{x} - \sigma$ and the other at $\bar{x} + \sigma$. The area under the curve between the two lines is proportional to the number of data lying between $\bar{x} - \sigma$ and $\bar{x} + \sigma$. It can be shown that almost 70% of the total area under the whole curve lies between $\pm\sigma$ of the mean indicating that $\approx$70% of the data lie between $\bar{x} - \sigma$ and $\bar{x} + \sigma$. In addition, $\approx$95% of data lie between $\bar{x} - 2\sigma$ and $\bar{x} + 2\sigma$.

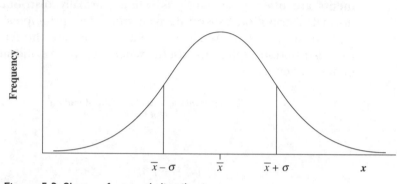

Figure 5.3: Shape of normal distribution

EXERCISE B

In an optics experiment, 60 repeat measurements of the distance between a mirror and an image were made. The measured values are shown in table 5.10.

Table 5.10: Values from 60 repeat measurements of an image position

Image distance (mm)					
146	174	170	165	175	167
163	171	161	166	157	173
162	148	180	165	175	176
158	174	161	176	166	158
164	169	157	153	175	159
159	160	174	156	166	166
173	158	168	166	173	159
151	160	155	168	166	165
164	175	161	171	167	166
164	163	156	162	167	164

The incomplete table following gives the number of data falling into the given intervals.

Table 5.11: Frequencies of experimental values falling within given intervals

Interval, x (mm)	Frequency
$145 \leq x < 150$	2
$150 \leq x < 155$	2
$155 \leq x < 160$	11
$160 \leq x < 165$	
$165 \leq x < 170$	
$170 \leq x < 175$	
$175 \leq x < 180$	
$180 \leq x < 185$	

1. Complete table 5.11 using the data in table 5.10.
2. Draw up a histogram using table 5.11. Do the data look to be normally distributed?
3. The mean and standard deviation of the data in table 5.10 are, $\bar{x}$ = 164.8 mm and σ = 7.3 mm. How many data values lie within $\pm\sigma$ of the mean and how many within $\pm2\sigma$ of the mean? Is this consistent with the normal distribution?

5.4.2 Population and sample

Although we must do all we can to ensure that the data we gather during an experiment are reliable, reproducible and have uncertainties reduced to as small a level as possible, these actions cannot extend to making an infinite number of repeat measurements of the same quantity. After all, though important, data gathering represents only one element of a well-planned and executed experiment. The totality of measurements that *could* be made is referred to as the *population* of measurements with mean, μ, and standard deviation, σ_{POP} given by:[8]

$$\mu = \frac{\sum x_i}{n} \qquad (5.5)$$

---

8. A number which is characteristic of the population of data is sometimes referred to as a population *parameter*.

$$\sigma_{POP} = \left(\frac{\sum (x_i - \mu)^2}{n} \right)^{\frac{1}{2}} \tag{5.6}$$

Where n is the totality of measurements that could be made.

We can take the population mean, μ, to be the true value of the quantity being sought when making a measurement. That this value can only be found after an infinite number of measurements have been made makes it of little practical use. What we *are* able to do is to make a few repeat measurements which can be regarded as a *sample* of all possible measurements and *estimate* the population mean (and standard deviation) using those values. The estimate of the population mean is simply the mean, $\bar{x}$, of all the data values.

The symbol, s, is used to represent the estimate of the population standard deviation and is calculated using an equation very similar to equation 5.2:

$$\sigma_{POP} \approx s = \left(\frac{\sum (x_i - \bar{x})^2}{n-1} \right)^{\frac{1}{2}} \tag{5.7}$$

Equation 5.7 allows us to estimate the population standard deviation using a sample taken from the population. This is in contrast to equation 5.2 which gives the standard deviation of the sample. (Appendix 2 discusses briefly why $n - 1$, rather than n, appears in equation 5.7.)

We want information concerning the population from which a sample is drawn therefore equation 5.7 is preferred to equation 5.2 when calculating the standard deviation. In practice, however, so long as the number of repeated measurements is greater than 3, equations 5.2 and 5.7 will usually return the same number to one significant figure. This is good enough for most uncertainty calculations involving the standard deviation. The larger the value of n, the closer will be the values produced by equations 5.2 and 5.7.

EXERCISE C

Consider the following 12 numbers.

86   70   75   78   90   59   78   82   81   70   70   69

(i) For the above numbers, calculate $\bar{x}$, σ, s.

(ii) Calculate $\sigma_{\bar{x}}$ using first σ then s as the standard deviation.

(iii) Give the mean and the uncertainty in the mean to the appropriate number of significant figures.

5.4.3 Confidence limits

The distribution of means of samples taken during an experiment follows a normal distribution in the same way as the 'raw' data. Figure 5.4 shows the distribution of means.

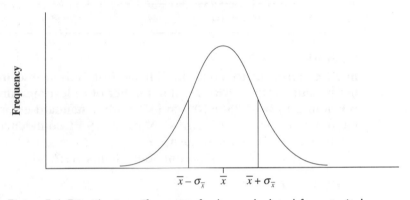

Figure 5.4: Distribution of means of values calculated from typical experimental data

The major difference between the distribution of the raw data and the distribution of the means is in the width of the distribution given by the standard deviation, σ, in the case of the individual measurements, and $\sigma_{\bar{x}} = \dfrac{\sigma}{\sqrt{n}}$, for the means, where n is the number of repeat measurements made.

We saw in section 5.4.1 that 70% of data values lie between $+\sigma$ of the mean and 95% between $\pm 2\sigma$. When dealing with the standard error of the mean we can interpret $\pm \sigma_{\bar{x}}$ as follows: *The true mean has ≈70% chance of lying between $\pm \sigma_{\bar{x}}$ of the mean of the sample and ≈95% chance of lying between $\pm 2\sigma_{\bar{x}}$ of the mean of the sample.*

We see that it *is* possible for the true value of the quantity being measured to be outside the limits $\bar{x} - 2\sigma_{\bar{x}}$ to $\bar{x} + 2\sigma_{\bar{x}}$. However, that probability is low at 5%, or one chance in twenty. We speak of *confidence limits* associated with the standard error of the mean. Table 5.12 (overleaf) shows a summary of the confidence limits and their associated probability.

Table 5.12: Confidence limits and the probability that the true value of a quantity lies within these limits

Confidence limits	Probability that true value lies between these limits (%)
$\bar{x} - \sigma_{\bar{x}}$ to $\bar{x} + \sigma_{\bar{x}}$	68.3
$\bar{x} - 2\sigma_{\bar{x}}$ to $\bar{x} + 2\sigma_{\bar{x}}$	95.4
$\bar{x} - 3\sigma_{\bar{x}}$ to $\bar{x} + 3\sigma_{\bar{x}}$	99.7
$\bar{x} - 4\sigma_{\bar{x}}$ to $\bar{x} + 4\sigma_{\bar{x}}$	99.994

Example

In an experiment to study the diffusion of carbon into iron, the mean value for diffusivity, $\bar{D}$, of a number of repeat measurements was found to be 2.75×10^{-7} cm^2 s^{-1}, with a standard error in the mean, $\sigma_{\bar{D}}$, of 0.07×10^{-7} cm^2 s^{-1}. What are 95% confidence limits for the true value of the diffusivity?

Using table 5.12, the 95% confidence limits are:[9]

$$\bar{D} - 2\sigma_{\bar{D}} \text{ to } \bar{D} + 2\sigma_{\bar{D}}$$

that is $(2.75 \times 10^{-7} - 2 \times 0.07 \times 10^{-7})$ cm^2 s^{-1} to
$(2.75 \times 10^{-7} + 2 \times 0.07 \times 10^{-7})$ cm^2 s^{-1}

$= 2.61 \times 10^{-7}$ cm^2 s^{-1} to 2.89×10^{-7} cm^2 s^{-1}

This can be written much more neatly as:

$$(2.75 \pm 0.14) \times 10^{-7} \text{ cm}^2 \text{ s}^{-1}$$

As for uncertainties, it is reasonable to give the number following the $\pm$ sign to one significant figure unless the first figure is a '1' (as in this example) in which case the number is given to two significant figures.

5.4.4 Review of uncertainty

It can sometimes be difficult to interpret the uncertainty given in an experiment. If, following a study on the mechanical properties of copper, the experimenter quotes the Young's modulus as $(0.95 \pm 0.05) \times 10^{11}$ Pa, do we take it that the true value for the

---

9. When the number of repeated measurements are small, the normal distribution tends to underestimate the confidence limits, and the 't' distribution is sometimes used (see appendix 2 for more details).

Young's modulus definitely lies between 0.90×10^{11} Pa and 1.00×10^{11} Pa? How has the uncertainty of 0.05×10^{11} Pa been calculated? If the details of the uncertainty calculation are not given, we are forced to make an educated guess as to what the uncertainty represents. The interpretation of the uncertainty is straightforward if it is given in terms of the standard error of the mean of the measurements made, and this method is preferred in most situations.

Whatever approach to presenting uncertainties is adopted, sufficient details should be given about the method by which the uncertainty was calculated so that a reader can interpret this number accordingly.

5.5 Combining uncertainties when the uncertainties in the measured quantities are independent

In chapter 4 we used a method for combining uncertainties which, though satisfactory in many situations, tends to overestimate the uncertainty in the calculated quantity. The problem is this: if quantity a has uncertainty Δa and quantity b has uncertainty Δb, it is possible for those uncertainties to partially cancel out in situations where the uncertainties in a and b are independent of each other. We are saying that sometimes Δa can be positive when Δb is negative, and vice versa, but that there is no dependence between the sign and size of Δa and the sign and size of Δb.

If a quantity V depends on the quantities a and b, we can write:

$$V = V(a, b)$$

In appendix 3 we show that the variance in the value of V, σ_V^2, is related to the variances in a and b, σ_a^2 and σ_b^2 respectively, in the following way (so long as the uncertainties in a and b are independent of each other):

$$\sigma_V^2 = \left(\frac{\partial V}{\partial a}\right)^2 \sigma_a^2 + \left(\frac{\partial V}{\partial b}\right)^2 \sigma_b^2 \qquad (5.8)$$

Taking the standard error as the uncertainty in the mean of the measured values of V, a and b we can rewrite equation 5.8 in terms of the standard errors $\sigma_{\bar{V}}$, $\sigma_{\bar{a}}$ and $\sigma_{\bar{b}}$:

$$\sigma_{\bar{V}} = \sqrt{\left(\frac{\partial V}{\partial a}\right)^2 \sigma_{\bar{a}}^2 + \left(\frac{\partial V}{\partial b}\right)^2 \sigma_{\bar{b}}^2} \qquad (5.9)$$

Example

In an experiment to establish the density of a particular metal, the mass and volume of a sample were measured a number of times. Table 5.13 shows the means and standard errors for the mass and volume measurements.

Table 5.13: Mean mass and volume of a metal sample. standard errors in these quantities are also shown.

Mean mass, $\overline{m}$ (g)	Standard error in $\overline{m}$, $\sigma_{\overline{m}}$ (g)	Mean volume, (mm^3)	Standard error in $\overline{V}$, $\sigma_{\overline{V}}$ (mm^3)
1.28	0.03	149	6

Given that the relationship between density, ρ, mass, m and volume, V is:

$$\rho = \frac{m}{V}$$

calculate the standard error in ρ using the values in table 5.13.

Answer

We can write equation 5.9 in terms of ρ, m and V:

$$\sigma_\rho = \sqrt{\left(\frac{\partial \rho}{\partial m}\right)^2 \sigma_m^2 + \left(\frac{\partial \rho}{\partial V}\right)^2 \sigma_V^2} \qquad (5.10)$$

$$\frac{\partial \rho}{\partial m} = \frac{1}{V}$$

$$= \frac{1}{149 \text{ mm}^3} = 6.711 \times 10^{-3} \text{ mm}^{-3}$$

and

$$\frac{\partial \rho}{\partial V} = -\frac{m}{V^2}$$

$$= -\frac{1.28 \text{ g}}{(149 \text{ mm}^3)^2} = -5.766 \times 10^{-5} \text{ g mm}^{-6}$$

Substituting these values into equation 5.9 gives:

$$\sigma_\rho = \sqrt{(6.711 \times 10^{-3} \text{ mm}^{-3})^2 \times (0.03 \text{ g})^2 + (-5.766 \times 10^{-5} \text{ g mm}^{-6})^2 \times (6 \text{ mm}^3)^2}$$

$$= \sqrt{1.602 \times 10^{-7} \text{ g}^2 \text{ mm}^{-6}}$$

$$= 4.003 \times 10^{-4} \text{ g mm}^{-3}$$

The density of the sample,

$$\rho = \frac{m}{V} = \frac{1.28 \text{ g}}{149 \text{ mm}^3} = 8.591 \times 10^{-3} \text{ g mm}^{-3}$$

The density and the uncertainty in the density can now be quoted to the correct number of significant figures as:

$$\rho = (8.6 \pm 0.4) \; 10^{-3} \text{ g mm}^{-3}$$

EXERCISE D

1. Using the data given in the previous example, calculate the uncertainty in ρ using the method dealing with the combination of uncertainties described in section 4.5.3.1.
2. In an experiment to find the refractive index, n, of a block of glass, the angle of incidence, i, and angle of refraction, r, were measured a number of times. Table 5.14 shows the mean of both quantities along with the standard errors.

Table 5.14: Means and standard errors of angles of incidence and refraction

Mean angle of incidence, $\bar{i}$ (°)	Standard error in $\bar{i}$, $\sigma_{\bar{i}}$ (°)	Mean angle of refraction, $\bar{r}$ (°)	Standard error in $\bar{r}$, $\sigma_{\bar{r}}$ (°)
61	1	38	2

Given that the relationship between n, i and r is:

$$n = \frac{\sin i}{\sin r}$$

(i) Calculate the best estimate for n.

(ii) Determine $\dfrac{\partial n}{\partial i}$ and $\dfrac{\partial n}{\partial r}$.

(iii) Write equation 5.8 in terms of n, i and r.

(iv) Use the equation in (iii) to find the standard error in n. (Note that your calculations require that the angles be expressed in radians, not degrees.)

(v) Give the best estimate for n and the standard error in n to the appropriate number of significant figures.

5.6 Continuous and discrete quantities

The quantities we have considered so far vary smoothly, or *continuously,* and the resolution of our measurements is only limited by the resolution capabilities of the particular instrument being used. Table 5.15 gives just a few examples of physical quantities that vary continuously.

Table 5.15: Examples of continuously varying quantities

Continuously varying quantities	time temperature length pressure mass force voltage

When quantities which vary continuously are measured experimentally, such as those shown in table 5.15, their variation can be described by a continuous distribution of which the normal distribution is the most widely used example.

However, there is an important class of experiments in which the quantity measured does *not* vary continuously, but increases by whole number amounts. We can refer to these as counting experiments. Situations which involve counting include X-ray or radioactivity experiments in which the number of X-rays or emitted particles are counted over a period of time.

We can refer to the detection of an X-ray or any other particle as an 'event'. If the probability of the event occurring in a small time interval is small and the occurrence of one event has no influence on later events, then the spread of the number of events occurring in a fixed time period can be described by a discrete distribution called the Poisson distribution.[10]

If the number of counts over a fixed-time period is recorded in a radioactivity experiment, and this process is repeated many times, a distribution of discrete data occurs which can be presented in the form of a histogram. Figure 5.5 shows a distribution of data for a radioactive counting experiment in which 500 repeat measurements of the number of counts recorded in a one-minute period were made.

Note the similarity in the shape of the above distribution to that of the normal distribution. In fact, in situations in which the number of counts recorded, N, is greater than 10 it is quite acceptable to use the normal distribution as an approximation to the Poisson distribution. Calculating the mean and the standard

---

10. For details of the Poisson distribution see chapter 11 of *An Introduction to Error Analysis* (full reference in appendix 1).

deviation of the distribution shown in figure 5.5 we find that the mean number of counts, $\overline{N}$, ≈ 440 and the standard deviation of the number of counts, σ, ≈ 21. This points us to a very simple relationship between $\overline{N}$ and σ for the Poisson distribution which can be justified theoretically. It is that:

$$\sigma = \sqrt{\overline{N}} \qquad\qquad (5.11)$$

If N is the number of counts recorded in a single measurement, then this value must be regarded as our 'best estimate' for the true value of the number of counts and the standard deviation of this value is $\approx \sqrt{N}$.

The standard error in the mean of n repeat measurements, $\sigma_{\overline{N}}$, is written as before (see equation 5.3):

$$\sigma_{\overline{N}} = \frac{\sigma}{\sqrt{n}} \qquad\qquad (5.12)$$

Substituting 5.11 into 5.12 gives:

$$\sigma_{\overline{N}} = \sqrt{\frac{\overline{N}}{n}} \qquad\qquad (5.13)$$

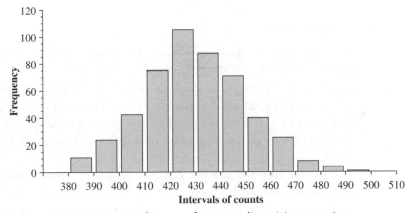

Frequency of counts as a function of count interval

Figure 5.5: Distribution of counts from a radioactivity counting experiment

Example

In an X-ray experiment, the number of X-rays from a powdered sample of nickel oxide over a period of 10 minutes was measured on ten occasions. Table 5.16 (overleaf) shows the number of counts.

Table 5.16: Ten repeat measurements of the number of X-rays emitted from a sample of nickel oxide over a period of 10 minutes

19	28	28	36	31	23	31	33	30	21

(i) Find the mean of the counts shown in table 5.16.

(ii) Calculate the standard deviation, σ, of the number of counts.

(iii) Calculate the standard error in the mean value.

Answer

(i) $\bar{N}$ = 28.0 counts.

(ii) Using equation 5.11, $\sigma = 5.29$.

(iii) We can use either equation 5.12 or 5.13 to calculate $\sigma_{\bar{N}}$. Using 5.12 we have:

$$\sigma_{\bar{N}} = \frac{\sigma}{\sqrt{n}} = \frac{5.29}{\sqrt{10}} = 1.67$$

so we can write the number of counts in the ten-minute period as:

$$N = 28.0 \pm 1.7$$

5.7 Comment

The statistical methods we have introduced in this chapter for dealing with the variability in data are very powerful and of wide applicability. Perhaps the greatest benefit to be derived from their use is in clarifying what is meant by the uncertainty in a quantity. Limits set by the standard error of the mean are directly related to the probability of the true value of a quantity lying between those limits.

Whether the method for combining uncertainties discussed in the chapter should be used, requires the assumption of independence of uncertainties to be valid. If this assumption is not valid, then the methods for combining uncertainties described in sections 4.5 to 4.5.4 are preferred. In situations where it is unclear whether the uncertainties are independent or not, then either method can be used. What *is* important is that whatever method is chosen, it should be described fully in your laboratory notebook or report.

PROBLEMS

1. Table 5.17 shows ten random numbers in the range 0 to 1.

Table 5.17: Ten random numbers in the range 0 to 1

0.549	0.022	0.295	0.178	0.190	0.425	0.672	0.996	0.573	0.934

   (i) Take the first two numbers in table 5.17 and calculate σ and s (given by equations 5.2 and 5.7 respectively). Give σ and s to two significant figures.
   (ii) Repeat (i) but this time calculate σ and s using the first three numbers in table 5.17, and then the first four numbers and so on. At what stage do σ and s agree to one significant figure?

2. A particular oxide loses oxygen when it is heated above 500°C in a vacuum. Table 5.18 shows the mass loss from twelve 500-mg samples of the oxide which were heated to a temperature of 800°C for 2 hours.

Table 5.18: Mass lost by 12 samples of a ceramic

Mass loss (mg)	9.5	9.9	8.6	8.8	8.8	9.0	8.6	9.1	9.5	8.3	9.6	9.6

Using the data in table 5.18 calculate:
   (i) the sample mean, variance and standard deviation
   (ii) the standard error of the mean
   (iii) the 70% confidence limits for true value of the mass loss.

3. A starting pistol is fired and a student standing (352 ± 5) m from the pistol measures the time elapsed between seeing the flash from the pistol and hearing the associated noise. Table 5.19 shows 50 consecutive measurements of the elapsed time.
   (i) Present the data in the form of a histogram.
   (ii) Do the data look to be normally distributed?
   (iii) Calculate the mean, estimate of population standard deviation and standard error of the mean.
   (iv) Given that the velocity of sound is the distance travelled divided by the elapsed time, calculate the velocity and the uncertainty in the velocity assuming uncertainties in distance and time measurements to be independent.

103

Table 5.19: 50 successive measurements of time

Time (s)				
1.24	0.70	1.02	1.07	0.87
1.07	0.87	1.28	1.23	1.10
0.90	1.24	0.82	1.02	1.35
0.96	1.03	1.09	1.31	1.59
1.04	1.07	1.09	1.13	1.36
0.87	1.29	1.34	1.42	0.89
1.53	1.06	1.58	0.98	1.01
1.43	0.80	1.18	1.00	0.74
0.99	0.95	0.97	0.85	1.22
1.10	1.28	1.18	1.16	0.84

4. Table 5.20 shows 20 successive measurements of the number of airborne particles in a fixed volume within a 'clean' room.

Table 5.20: Number of airborne particles in a fixed volume of air

153	132	143	152
159	136	160	165
158	122	149	138
169	170	144	161
147	149	162	150

Using the data in the table, calculate the 95% confidence limits for the true number of particles.

5. The optical density (o.d.) of a liquid is given by:

$$\text{o.d.} = \varepsilon C l$$

where ε is called the extinction coefficient, C is the concentration of the absorbing species in the liquid, and l is the path length of the light. Assuming that the uncertainties in ε, C and l are independent and that:

$$\varepsilon = (15 \pm 1) \text{ L mol}^{-1} \text{ mm}^{-1}$$

$$C = (0.04 \pm 0.01) \text{ mol L}^{-1}$$

$$l = (1.4 \pm 0.2) \text{ mm}$$

calculate o.d. and the uncertainty in this quantity.

6. The input voltage, V_{in}, to a particular amplifier can be calculated using the formula:

$$V_{in} = \frac{V_o R_f}{R_f + 39 \times 10^3}$$

where V_o is the output voltage from the amplifier and R_f is the value of a feedback resistor.

If $R_f = (1.00 \pm 0.05) \times 10^3 \ \Omega$ and $V_o = (253 \pm 5)$ mV, calculate V_{in} and the uncertainty in V_{in} assuming that the uncertainties in R_f and V_{in} are independent.

7. The viscosity of a fluid can be determined by studying the motion of a sphere falling through the liquid. A formula from which the viscosity of fluid, η_f can be calculated is:

$$\eta_f = \frac{\frac{2}{9} r^2 g (\rho_s - \rho_f)}{v_t}$$

where:

r is the radius of the sphere $= (1.2 \pm 0.1)$ mm

g is the acceleration due to gravity $= 9.81$ m s^{-2}

ρ_s is the density of the sphere $= 8615$ kg m^{-3}

ρ_f is the density of the fluid $= 1015$ kg m^{-3}

v_t is the terminal velocity of the sphere through the fluid $= (15 \pm 2)$ mm s^{-1}

Use this information to calculate the viscosity of the fluid and the uncertainty in the viscosity, assuming uncertainties to be independent.

6 FITTING A LINE TO *X-Y* DATA: LEAST SQUARES METHOD

6.1 Overview: how can we find the 'best' line through a set of points?

Linearly related *x-y* data occur so frequently in experiments in science and engineering that the analysis of such data deserves special attention. In general we seek to find the quantitative relationship which best describes the dependence of *y* upon *x*. In order to do this we need a method by which we can determine the equation of the line that best 'fits' the *x-y* data.

In chapter 3 we saw that an equation representing the relationship between *x* and *y* quantities can be found by first plotting the data on an *x-y* graph followed by drawing the 'best' line through the points (or at least as close as possible to them) with a rule. The gradient, *m*, and intercept, *c*, of this line are calculated and the equation of the line is written, $y = mx + c$.

Although positioning a line 'by eye' through the data points is a good way of obtaining reasonable estimates for *m* and *c*, there are a number of difficulties associated with this method:

(i) No two people draw the same 'best' line through a given data set.

(ii) If the uncertainty in each data point is different, how do we take this into account when drawing a line through the points?

(iii) Drawing the best line is difficult if the data exhibit large scatter.

(iv) Finding the uncertainties in *m* and *c* directly from the graph (as described in section 3.3.4.2) is cumbersome, and tends to overestimate their values.

Figure 6.1 shows an example of a graph in which it is difficult to determine the best line through the points.

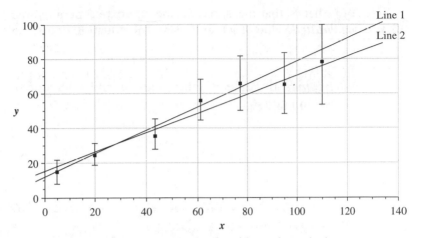

Figure 6.1: Linear x-y graph with two 'best' lines through the points

Both lines in figure 6.1 appear to fit the data quite well, but which is the better?[1] To answer this we need a new tool that will avoid the guesswork involved when finding the best line through a set of points by eye. That tool is usually described as the fitting of a line to data using the method of 'least squares' (also often referred to as *linear regression*).

6.2 The method of least squares

To find the best line through a set of data we begin by assuming that any random uncertainty in data values is confined to measurements made of the y-quantity. The assumption is often valid as it is the x-quantity that is controlled or adjusted in a stepwise fashion during an experiment and so it is usually possible to know this quantity to high precision. Secondly, we assume that the uncertainty in each measurement of the y-quantity is the same, which is equivalent to saying that the error bars attached to each point are of the same length. We will deal with the more general situation in which the uncertainties in the y-values vary from point to point in section 6.3.

Figure 6.2 shows part of an x-y graph with a line passing close to the data points. For a particular value of x, labelled x_i, there are two values of y shown on the graph: y_{io} refers to the *observed* value

---

1. For comparison, the gradient and intercept of line 1 are 0.68 and 12, respectively, while for line 2, the gradient and intercept are 0.55 and 16, respectively.

of y, that is, that measured during the experiment; y_{ic} refers to the *calculated* value of y found using the equation of a straight line:

$$y_{ic} = mx_i + c \qquad (6.1)$$

where m and c refer to the gradient and intercept of the line shown in figure 6.2.

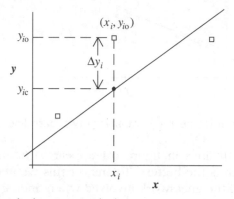

Figure 6.2: x-y graph showing residual, Δy_i

Δy_i is the difference between the observed and calculated y-value and is termed the *residual*, given by:

$$\Delta y_i = y_{io} - y_{ic} \qquad (6.2)$$

As we move the line around in an effort to find the position where the line passes closest to the majority of points, Δy_i for each point changes. A criterion is required by which we can decide the best position for the line. This position, and therefore the best values for m and c, is found by applying a theory from statistics called the *principle of maximum likelihood*.[2] This predicts that the best line will be found by minimising the sum of the *square* of the residuals. Writing the sum of squares as SS, we can say:

$$SS = (\Delta y_1)^2 + (\Delta y_2)^2 + (\Delta y_3)^2 + ... + (\Delta y_n)^2$$

---

2. The principle of maximum likelihood considers the probability of obtaining the observed values of x and y during an experiment and asserts that a particular combination of m and c values describes the linear relationship which caused those values to arise. By finding the values of m and c, which will make the probability of obtaining the observed set of y-values a maximum, the 'best' values of m and c are obtained. For more details refer to chapter 8 of *An Introduction to Error Analysis* (reference in appendix 1).

which can be abbreviated to:

$$SS = \sum_{i=1}^{t=n} (\Delta y_i)^2 \qquad (6.3)$$

The summation indicates that the square of the residuals must be added up for all the data points from $i = 1$ to $i = n$ where n is the number of points. To make future equations less cumbersome, we will leave out the limits of the summation and assume that all sums are calculated from $i = 1$ to $i = n$.

Replacing Δy_i in equation 6.3 by $y_{io} - y_{ic}$ and y_{ic} by $mx_i + c$ we can write:

$$SS = \sum (y_{io} - (mx_i + c))^2 \qquad (6.4)$$

As already stated, we must find values for m and c that reduce SS to the smallest possible value. Those values are taken to be the best values for gradient and intercept. Mathematics gives us the tool for finding values for m and c which minimise SS. First, equation 6.4 is partially differentiated with respect to m and the resulting equation set equal to zero. The next step is to partially differentiate equation 6.4 with respect to c and set the resulting equation equal to zero. Doing this gives:

$$\sum [x_i (y_{io} - mx_i - c)] = 0 \qquad (6.5)$$

and:

$$\sum (y_{io} - mx_i - c) = 0 \qquad (6.6)$$

Equations 6.5 and 6.6 can be expanded and combined to give the following equations[3] for m and c:

$$m = \frac{n\sum x_i y_i - \sum x_i \sum y_i}{n\sum x_i^2 - (\sum x_i)^2} \qquad (6.7)$$

and

$$c = \frac{\sum x_i^2 \sum y_i - \sum x_i \sum x_i y_i}{n\sum x_i^2 - (\sum x_i)^2} \qquad (6.8)$$

---

3. It is usual to drop the 'o' from the subscript of y and to take it as read that it is the observed values of y that appear in the equations.

Equations 6.7 and 6.8 are extremely useful. In situations where the uncertainties in the measured *y*-values do not vary from measurement to measurement, and where there are negligible uncertainties in the *x*-values, they allow us to calculate the best values for the gradient, *m* and the intercept, *c*.

Let us consider an example of the use of equations 6.7 and 6.8.

6.2.1 Example of fitting a line to *x*-*y* data

In an experiment to study the behaviour of silicon diodes when cooled, the voltage across a diode was measured as a function of the diode temperature. Figure 6.3 shows a graph of the data gathered.

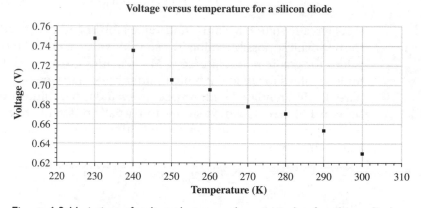

Figure 6.3: Variation of voltage between the terminals of a silicon diode with temperature

Over the range of temperature shown in figure 6.3, the relationship between voltage and temperature appears to be linear, therefore we will use equations 6.7 and 6.8 to find the best line through the points. We begin by assigning the temperature as the *x*-quantity and the voltage as the *y*-quantity. The next step is to draw up a table so that the quantities $\sum x_i$, $\sum y_i$, $\sum x_i y_i$ and $\sum x_i^2$ can be calculated. Table 6.1 contains the data (with units included in the column headings) shown plotted in figure 6.3.

Now use equation 6.7 to find the gradient of the line:

$$m = \frac{8 \times 1454.42 \text{ K V} - 2120 \text{ K} \times 5.514 \text{ V}}{8 \times 566\,600 \text{ K}^2 - (2120 \text{ K})^2} = \frac{-54.32 \text{ K V}}{33\,600 \text{ K}^2}$$

$$= -1.61\dot{6} \times 10^{-3} \text{V K}^{-1}$$

and use equation 6.8 to find the intercept of the line on the y-axis, c:

$$c = \frac{566\,000\ \text{K}^2 \times 5.514\ \text{V} - 2120\ \text{K} \times 1454.42\ \text{K V}}{8 \times 566\,000\ \text{K}^2 - (2120\ \text{K})^2} = \frac{37\,553.6\ \text{K}^2\ \text{V}}{33\,600\ \text{K}^2}$$

$$= 1.117\dot{6}\ \text{V}$$

Table 6.1: Columns required for fitting a line to data using the method of least squares

x_i (K)	y_i (V)	$x_i y_i$ (K V)	x_i^2 (K²)
300	0.630	189.00	90 000
290	0.653	189.37	84 100
280	0.670	187.60	78 400
270	0.678	183.06	72 900
260	0.695	180.70	67 600
250	0.705	176.25	62 500
240	0.735	176.40	57 600
230	0.748	172.04	52 900
$\sum x_i = 2120$	$\sum y_i = 5.514$	$\sum x_i y_i = 1454.42$	$\sum(x_i^2) = 566\,000$

Figure 6.4 shows the best line through the points using the gradient and intercept just calculated.

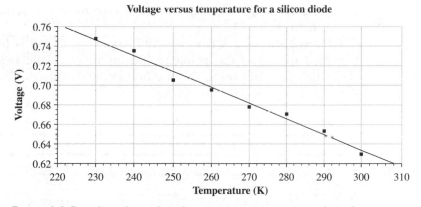

Figure 6.4: Best line through voltage versus temperature data shown in table 6.1

6.2.2 Possible calculation difficulties

When finding m and c using equations 6.7 and 6.8, care must be taken not to round intermediate values of the calculations as this can influence the values of m and c greatly. For example, the numerator in the calculation for m in the example above is -54.32 K V. If each value in the numerator is rounded to three significant figures (for example 1454.42 K V is rounded to 1450 K V) then the numerator turns out to be -81.2 K V. Rounding has a big effect when terms are subtracted in the numerator (or the denominator) which are about the same size, as in this example. It is good advice to keep any intermediate results to as many figures as possible and only round values at the end. The problem is minimised if fitting is performed using built-in least squares functions on a pocket calculator (this is dealt with in chapter 8). A calculator normally holds numbers to a precision of about 12 figures and so the influence of rounding on the final values of m and c is minimal.

EXERCISE A

1. In order to find the gradient of a particular straight line, the following calculation was performed:

$$m = \frac{10 \times 672.93 - 2350.6 \times 2.785}{10 \times 552\,900 - (2350.6)^2}$$

  (i) Calculate m using the numbers as shown. Give m to two significant figures.
  (ii) Round each number shown above to *three* significant figures and recalculate m. Again give m to two significant figures.

2. Table 6.2 shows a set of x-y data. Assuming that y is linearly related to x, find the gradient and intercept of the best line through the data using the method of least squares. Give the values to four significant figures.

Table 6.2: Linearly related x-y data

x	y
2.1	45.4
4.4	65.7
6.3	73.4
8.3	95.0

Table 6.2 (continued)

x	y
10.2	102.8
12.3	121.2
14.6	134.7
16.7	155.3

3. The acceleration due to gravity, g', was measured at various heights, h, above the Earth's surface. Table 6.3 gives the values of g' corresponding to various heights.

Table 6.3: Variation of acceleration due to gravity with height above the earth's surface

h (m)	g' (m s^{-2})
1×10^4	9.76
2×10^4	9.74
3×10^4	9.70
4×10^4	9.69
5×10^4	9.66
6×10^4	9.62
7×10^4	9.59
8×10^4	9.55
9×10^4	9.54
10×10^4	9.51

(i) Taking height as the independent (x) variable, and the acceleration due to gravity as the dependent variable (y), calculate the gradient and intercept using the least squares method.

(ii) The approximate relationship between g' and h is:

$$g' = g\left(1 - \frac{2h}{R_E}\right)$$

where R_E is the radius of the Earth and g is the acceleration due to gravity on the surface of the Earth. Rearrange the above equation into the form $y = mx + c$, and use your values for m and c to find values for R_E and g.

6.2.3 Uncertainty in gradient and intercept

We can find the best values for the gradient and intercept of a line through a set of x-y data using equations 6.7 and 6.8. However, it is not possible to decide how many figures m and c should be quoted to until the uncertainties in m and c, which we will write as σ_m and σ_c respectively, have been calculated.[4]

In order to calculate σ_m and σ_c we assume the following:

(i) For each value of x, the corresponding value of y has some uncertainty.

(ii) The uncertainty in each value of y contributes something to the uncertainties in m and c.

(iii) If the uncertainties in y are independent, we can adapt equation 5.9 to give the uncertainty in m as:

$$\sigma_m = \sqrt{\left(\frac{\partial m}{\partial y_1}\right)^2 \sigma_1^2 + \left(\frac{\partial m}{\partial y_2}\right)^2 \sigma_2^2 + \ldots\ldots + \left(\frac{\partial m}{\partial y_n}\right)^2 \sigma_n^2} \quad (6.9)$$

where $\sigma_1, \sigma_2 \ldots \sigma_n$ are the uncertainties in the observed values $y_1, y_2 \ldots y_n$. If the uncertainties in y are constant, then $\sigma_1, \sigma_2 \ldots \sigma_n$ can each be replaced by σ allowing equation 6.9 to be written:

$$\sigma_m = \left[\sigma^2 \sum_{i=1}^{i=n}\left(\frac{\partial m}{\partial y_i}\right)^2\right]^{\frac{1}{2}} \quad (6.10)$$

A similar equation to 6.10 can be written for the variance in σ_c by simply replacing m in the equation by c:

$$\sigma_c = \left[\sigma^2 \sum_{i=1}^{i=n}\left(\frac{\partial c}{\partial y_i}\right)^2\right]^{\frac{1}{2}} \quad (6.11)$$

There are a number of mathematical steps required before we can arrive at explicit equations for σ_m and σ_c. We will not go through the steps here, but instead simply quote the results:

$$\sigma_m = \frac{\sigma n^{\frac{1}{2}}}{\left[n \sum x_i^2 - \left(\sum x_i\right)^2\right]^{\frac{1}{2}}} \quad (6.12)$$

$$\sigma_c = \frac{\sigma \left(\sum x_i^2\right)^{\frac{1}{2}}}{\left[n \sum x_i^2 - \left(\sum x_i\right)^2\right]^{\frac{1}{2}}} \quad (6.13)$$

---

4. σ_m and σ_c are the standard errors in m and c and are taken to be the uncertainties in these quantities.

where σ is the uncertainty in each y-value of the data point. It is usual, when fitting a line to data in which the uncertainty in each point is constant, to take this uncertainty to be the standard deviation of the distribution of the y-values about the fitted line. This is given by:

$$\sigma = \left[\frac{1}{n-2} \sum (y_i - mx_i - c)^2 \right]^{\frac{1}{2}} \qquad (6.14)$$

The formula for σ is similar to the approach taken in equation 5.7 for estimating the standard deviation of a population based on a sample of data drawn from that population.[5] The deviation of each point with respect to the line is $y_i - mx_i - c$. The deviation is squared and summed for all data points. The reason that $n-2$ appears in the denominator of equation 6.14 is discussed briefly in appendix 2.

6.2.4 Example of a calculation for finding the uncertainty in m and c

We will use equations 6.12 to 6.14 to calculate the uncertainty in m and c for the data given in section 6.2.1. Fitting by least squares gave m and c as,

$$m = -1.616 \times 10^{-3} \, \text{V K}^{-1}$$

$$c = 1.1176 \, \text{V}$$

A new table is drawn up so that σ can be calculated.

Table 6.4: Columns required for calculation of uncertainty in m and c

x_i (K)	y_i (V)	$y_i - mx_i - c$ (V)	$(y_i - mx_i - c)^2$ (V²)
300	0.630	-2.6667×10^{-3}	7.1111×10^{-6}
290	0.653	4.1667×10^{-3}	1.7361×10^{-5}
280	0.670	5.0000×10^{-3}	2.5000×10^{-5}
270	0.678	-3.1667×10^{-3}	1.0028×10^{-5}
260	0.695	-2.3333×10^{-3}	5.4444×10^{-6}
250	0.705	-8.5000×10^{-3}	7.2250×10^{-5}
240	0.735	5.3333×10^{-3}	2.8444×10^{-5}
230	0.748	2.1667×10^{-3}	4.6944×10^{-6}
		$\sum (y_i - mx_i - c)^2 = 1.7033 \times 10^{-4}$	

5. For a derivation of equation 6.14, refer to chapter 6 of *Data Reduction and Error Analysis for the Physical Sciences* (full details given in appendix 1).

Using equation 6.14:

$$\sigma = \left(\frac{1.7033 \times 10^{-4} \; V^2}{8-2} \right)^{\frac{1}{2}} = 5.3281 \times 10^{-3} \, V$$

σ_m and σ_c are calculated using equations 6.12 and 6.13. $n \sum x_i^2 - \left(\sum x_i \right)^2$ was calculated for this data set in section 6.2.1 and found to be equal to $33\,600 \; K^2$. It follows that:

$$\sigma_m = \frac{5.3281 \times 10^{-3} \; V \times 8^{\frac{1}{2}}}{\left(33\,600 \; K^2 \right)^{\frac{1}{2}}} = 8.2214 \times 10^{-5} \, V \, K^{-1}$$

To calculate σ_c, we need $\sum x_i^2$ which is given as $566\,000$ in table 6.1. It follows that:

$$\sigma_c = \frac{5.3281 \times 10^{-3} \; V \times \left(566\,000 \; K^2 \right)^{\frac{1}{2}}}{\left(33\,600 \; K^2 \right)^{\frac{1}{2}}} = 0.021868 \, V$$

We are now in a position to quote m and c to a number of figures consistent with the uncertainties in each of these quantities.

$$m = (-1.62 \pm 0.08) \times 10^{-3} \, V \, K^{-1}, \text{and}$$

$$c = (1.12 \pm 0.02) \, V$$

As usual, the uncertainties have been rounded to one significant figure.

EXERCISE B

Refer to questions 2 and 3 of exercise A. Calculate the uncertainty in the gradient and intercept for the data given in each question. Quote the gradient and intercept along with the uncertainties to an appropriate number of significant figures.

6.2.5 Interpretation of uncertainties in m and c

σ_m and σ_c, which we take to be the uncertainties in m and c, are the standard errors of these quantities. We saw in section 5.4.3 that the true (or population) mean of a distribution of numbers, has about a 70% chance of lying within one standard error of the sample mean and about a 95% chance of lying within two standard errors of the sample mean. Similarly, when quoting σ_m, we are saying that the true value of the gradient has a 70% chance of lying within $\pm \sigma_m$ of m. The same argument holds for σ_c.

6.3 Weighting the fit

To take into account situations in which the uncertainties in the y-values vary from point to point, we use 'weighted' least squares when fitting a line to data. The sum of squares is weighted so that, when fitting takes place, the calculated line lies closest to those points that are known to the greatest precision, in effect 'favouring' those points. Situations in which a weighted fit is required are quite common and include where:

(i) changes in resolution occur while making measurements with an instrument which is switched between operating ranges during the course of an experiment. For example, a voltmeter might be used during part of an experiment on its 200-mV range with a corresponding resolution uncertainty of 0.1 mV. If later the voltmeter is switched to the 2-V range, the resolution uncertainty becomes 1 mV.

(ii) multiple measurements of y-values are made at some values of x and not at others. This results in error bars which are smaller for repeated measurements than for those where a 'one-off' measurement has been made.

(iii) y-values must be transformed so that a straight-line relationship can be produced.

Figure 6.5 shows linearly related data in which the uncertainties in the y-values are not constant. Whatever the reason for the variation in the size of the uncertainties, we must make the most of those points that have least uncertainty, and weight those with large uncertainty much less.

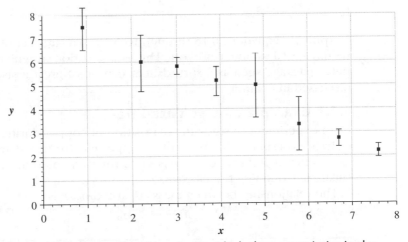

Figure 6.5: Linearly related x-y data in which the uncertainties in the y-values vary

When uncertainties vary from point to point, each value of uncertainty (written as σ_i) must be used in the calculations of m, σ_m, c, and σ_c.

In order to be able to write the equations for m, c etc. in a more condensed form, we introduce Δ, which is given by:

$$\Delta = \sum \frac{1}{\sigma_i^2} \sum \frac{x_i^2}{\sigma_i^2} - \left(\sum \frac{x_i}{\sigma_i^2} \right)^2 \tag{6.15}$$

The remaining quantities can be written as:[6]

$$m = \frac{\sum \dfrac{1}{\sigma_i^2} \sum \dfrac{x_i y_i}{\sigma_i^2} - \sum \dfrac{x_i}{\sigma_i^2} \sum \dfrac{y_i}{\sigma_i^2}}{\Delta} \tag{6.16}$$

$$\sigma_m = \left(\frac{\sum \dfrac{1}{\sigma_i^2}}{\Delta} \right)^{\frac{1}{2}} \tag{6.17}$$

$$c = \frac{\sum \dfrac{x_i^2}{\sigma_i^2} \sum \dfrac{y_i}{\sigma_i^2} - \sum \dfrac{x_i}{\sigma_i^2} \sum \dfrac{x_i y_i}{\sigma_i^2}}{\Delta} \tag{6.18}$$

$$\sigma_c = \left(\frac{\sum \dfrac{x_i^2}{\sigma_i^2}}{\Delta} \right)^{\frac{1}{2}} \tag{6.19}$$

Applying equations 6.15 through to 6.19 to a large set of x-y data requires a fair amount of work. This is a situation in which a computer package such as a spreadsheet can be of great assistance, as discussed in chapter 9.

6.3.1 Example of a weighted fit

In an experiment to study the deposition of copper during an electrolysis experiment, the mass of a copper electrode was measured as an electrical current was passed through a solution of copper sulfate surrounding the electrode.

The relationship between mass, M, and time, t, is:

$$M = kt + M_0$$

---

6. See chapter 6 of *Data Reduction and Error Analysis for the Physical Sciences* (see appendix 1) for a derivation of the equations 6.15 to 6.19.

where M_0 is the mass of the copper electrode at $t = 0$, and k is a constant.

When the electrode mass was below 50 g a balance was used, capable of resolving down to ± 0.1 g. When the mass exceeded 50 g, a balance with a larger range was used with a resolution of ± 0.5 g. Table 6.5 shows data gathered of mass, M, as a function of time, t, and the corresponding graph is shown in figure 6.6. As the error bars are not constant, a weighted fit is required.

Table 6.5: Variation of the mass of a copper electrode with time during an electrolysis experiment

t (min)	M (g)
10	47.7 ± 0.1
60	48.7 ± 0.1
90	49.4 ± 0.1
160	50.4 ± 0.5
200	51.6 ± 0.5
240	52.1 ± 0.5
300	52.7 ± 0.5

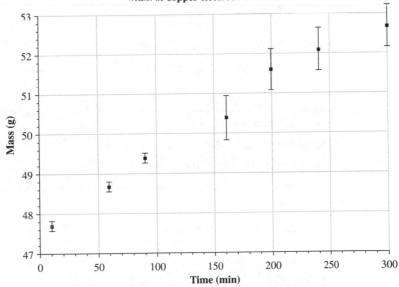

Figure 6.6: Graph of variation of the mass of a copper electrode with time during an electrolysis experiment

To assist in using equations 6.15 to 6.19, table 6.6 has been drawn up containing all the relevant quantities.

Table 6.6: Weighted fit of data shown

x_i (min)	y_i (g)	σ_i (g)	$\dfrac{1}{\sigma_i^2}$ (g^{-2})	$\dfrac{x_i}{\sigma_i^2}$ (min g^{-2})	$\dfrac{y_i}{\sigma_i^2}$ (g^{-1})	$\dfrac{x_i y_i}{\sigma_i^2}$ (min g^{-1})	$\dfrac{x_i^2}{\sigma_i^2}$ (min^2 g^{-2})
10	47.7	0.1	100	1000	4770	47 700	10 000
60	48.7	0.1	100	6000	4870	292 200	360 000
90	49.4	0.1	100	9000	4940	444 600	810 000
160	50.4	0.5	4	640	201.6	32 256	102 400
200	51.6	0.5	4	800	206.4	41 280	160 000
240	52.1	0.5	4	960	208.4	50 016	230 400
300	52.7	0.5	4	1200	210.8	63 240	360 000

$$\sum \frac{1}{\sigma_i^2} = 316 \qquad \sum \frac{x_i}{\sigma_i^2} = 19\,600 \qquad \sum \frac{y_i}{\sigma_i^2} = 15\,407.2 \qquad \sum \frac{x_i y_i}{\sigma_i^2} = 971\,292 \qquad \sum \frac{x_i^2}{\sigma_i^2} = 2\,032\,800$$

We can now calculate Δ using equation 6.15:

$$\Delta = 316 \text{ g}^{-2} \times 2\,032\,800 \text{ min}^2 \text{ g}^{-2} - (19\,600 \text{ min g}^{-2})^2$$
$$= 258\,204\,800 \text{ min}^2 \text{ g}^{-4}$$

m can be found using equation 6.16:

$$m = \frac{316 \text{ g}^{-2} \times 971\,292 \text{ min g}^{-1} - 19\,600 \text{ min g}^{-2} \times 15\,407.2 \text{ g}^{-1}}{258\,204\,800 \text{ min}^2 \text{ g}^{-4}} = \frac{4\,947\,152 \text{ min g}^{-3}}{258\,204\,800 \text{ min}^2 \text{ g}^{-4}}$$
$$= 1.9160 \times 10^{-2} \text{ g min}^{-1}$$

σ_m can be found using equation 6.17:

$$\sigma_m = \left(\frac{316 \text{ g}^{-2}}{258\,204\,800 \text{ min}^2 \text{ g}^{-4}} \right)^{\frac{1}{2}} = 1.11 \times 10^{-3} \text{ g min}^{-1}$$

The gradient is finally written as:

$$m = (1.92 \pm 0.11) \times 10^{-2} \text{ g min}^{-1}$$

Similarly, c and σ_c can be found using equations 6.18 and 6.19 respectively:

$$c = \frac{2\,032\,800 \text{ min}^2 \text{ g}^{-2} \times 15\,407.2 \text{ g}^{-1} - 19\,600 \text{ min g}^{-2} \times 971\,292 \text{ min g}^{-1}}{258\,204\,800 \text{ min}^2 \text{ g}^{-4}} = 47.5686 \text{ g}$$

$$\sigma_c = \left(\frac{2\,032\,800 \text{ min}^2 \text{ g}^{-2}}{258\,204\,800 \text{ min}^2 \text{ g}^{-4}} \right)^{\frac{1}{2}} = 8.87 \times 10^{-2} \text{ g}$$

so that c can be written as $c = (47.57 \pm 0.09)$ g.

A line of best fit can now be added to the data as shown in figure 6.7. For comparison, this graph shows the line of best fit using an unweighted fit.

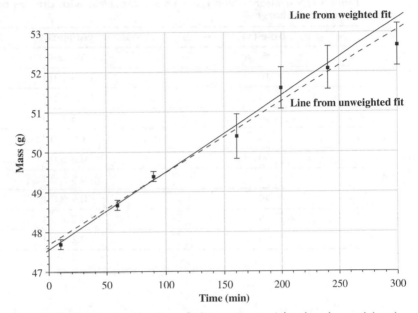

Figure 6.7: Graph showing best fit lines using weighted and unweighted least squares

Perform an *unweighted* fit to data shown in table 6.5 and show that the gradient and intercept created using this method are $m = (1.78 \pm 0.10) \times 10^{-2}$ g min^{-1} and $c = (47.67 \pm 0.18)$ g.

6.3.2 Example of a weighted fit where linearisation is required

In the previous example, the equation relating the data is of the form $y = mx + c$ so that no linearisation is required. When linearisation *is* required, it usually means that a weighted fit is also required even if the raw experimental data have uncertainties which are constant. Take, for example, an experiment in which the voltage across a capacitor is measured as the capacitor is discharged through a resistor. Table 6.7 shows the data of voltage and time gathered during the experiment. Figure 6.8 shows a plot of the data. (As the error bars are too small to show clearly, they have been omitted from the graph.)

Table 6.7: Variation of voltage across a capacitor with time as the capacitor discharges

time (s)	voltage (V)
5	10.9 ± 0.1
10	6.5 ± 0.1
15	4.1 ± 0.1
20	2.3 ± 0.1
25	1.5 ± 0.1
30	0.8 ± 0.1
35	0.5 ± 0.1
40	0.4 ± 0.1
45	0.2 ± 0.1
50	0.2 ± 0.1

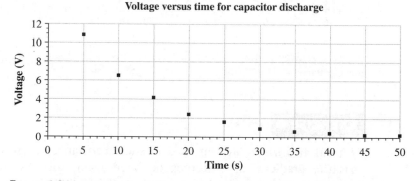

Voltage versus time for capacitor discharge

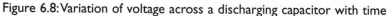

Figure 6.8: Variation of voltage across a discharging capacitor with time

The relationship between the voltage across a capacitor, V, and the time, t, as the capacitor discharges is:

$$V = V_0 \exp\left(\frac{-t}{\tau}\right) \tag{6.20}$$

where V_0 is the voltage at time $t = 0$ and τ is called the time constant for the discharge. Note that in table 6.7 the uncertainty in the voltage measurements is constant at 0.1 V. However, equation 6.20 must be transformed before a straight line can be fitted to the data and the uncertainty in the transformed quantity is *not* constant.

To transform equation 6.20 into an equation of the form $y = mx + c$, we take the natural logarithms of both sides of the equation. This gives:

$$\ln V = \ln V_0 - \frac{t}{\tau} \qquad (6.21)$$

Plotting $\ln V$ against t should give a straight line with a gradient of $-1/\tau$ and an intercept of $\ln V_0$. The quantity plotted on the y-axis is $\ln V$. If the uncertainty in V is 0.1 V, what is the uncertainty in $\ln V$? To answer this we use the mathematical tool of partial differentiation as introduced in chapter 4. If

$$y = f(V), \text{ then } \Delta y = \frac{\partial y}{\partial V} \Delta V$$

In this example, $y = \ln V$, so $\dfrac{\partial y}{\partial V} = \dfrac{1}{V}$

it follows that, $\Delta y = \dfrac{\Delta V}{V}$.

In this chapter we have used the symbol σ_i to represent the uncertainty in the ith y-value, so that in this example:

$$\sigma_i = \frac{\Delta V_i}{V_i}$$

Now we can draw up a table similar to table 6.6 so that the gradient and intercept of the $\ln V$ versus t data using weighted least squares can be found.

EXERCISE D

1. Table 6.8 shows the beginning of a table similar to table 6.6. The raw data is given in table 6.7.[7]

   Table 6.8: Columns required for the calculation of gradient and intercept (and uncertainties in these quantities) for the data given in table 6.7

x_i	$y_i = \ln V_i$	$\sigma_i = \dfrac{\Delta V_i}{V_i}$	$\dfrac{1}{\sigma_i^2}$	$\dfrac{x_i}{\sigma_i^2}$	$\dfrac{y_i}{\sigma_i^2}$	$\dfrac{x_i y_i}{\sigma_i^2}$	$\dfrac{x_i^2}{\sigma_i^2}$
(s)				(s)		(s)	(s^2)
5	2.388 763	0.009 174	11 881	59 405	28 380.89	141 904.5	297 025
10	1.871 802	0.015 385	4 225	42 250	7 908.364	79 083.64	422 500
15	1.410 987	0.024 39	1 681	25 215	2 371.869	35 578.04	378 225

7. Note that both y_i and σ_i have no units.

(i) Complete table 6.8 using the data in table 6.7.
(ii) Plot a graph of ln V versus t using the data shown in table 6.8.
(iii) Calculate the gradient and intercept of the line.
(iv) Calculate the uncertainties in the gradient and intercept.
(v) Find V_0 and τ and the associated uncertainties in these quantities.

2. In an experiment to study the rate of decay of a radioactive element, the following data were gathered.

Table 6.9: Variation of counts per second with time for radiation emitted from a radioactive source

t (s)	I (counts s^{-1})
60	1580
120	860
180	460
240	290
300	130
360	67
420	53
480	16

The relationship between number of counts per second, I, and time, t is:

$$I = I_0 \exp(-\lambda t)$$

where I_0 is the number of counts per second at time $t = 0$, and λ is decay constant of the radioactive element.

(i) Linearise the equation and plot the linearised data.
(ii) Taking the uncertainty in I to be equal to $\sqrt{I}$, do a weighted fit to data to find I_0 and λ and the uncertainties in these quantities.
(iii) Show the line of best fit on the graph.

3. In an optics experiment, the diameter of interference rings, D, was measured for each ring beginning close to the centre of the interference pattern. Table 6.10 shows the results obtained.

Table 6.10: Diameter of interference rings as a function of ring number

n	D (mm)
10	2.30 ± 0.02
11	2.35 ± 0.02
12	2.42 ± 0.02
13	2.50 ± 0.02
14	2.55 ± 0.02
15	2.61 ± 0.02
16	2.64 ± 0.02
17	2.71 ± 0.02
18	2.75 ± 0.02

The relationship between diameter, D, and ring number, n, is given by:

$$D = \sqrt{An + B}$$

where A and B are constants.
  (i) Linearise the above equation.
  (ii) Perform a least-squares fit to data to find A and B and the uncertainties in A and B.

6.4 Comment

This chapter has introduced the technique of fitting lines by least squares to linearly related x-y data. The technique can be extended to situations where equations have more than two parameters such as:

$$y = A \ln x + Bx + C$$

where A, B and C are the parameters. Appendix 1 gives details of books that can be referred to for details on how the least-squares approach can be applied in such situations.

In this chapter, discussion has been confined to cases where there is a linear relationship between x and y and the uncertainties in measured quantities are limited to the y-quantity. The least-squares functions found on calculators and in computer packages adopt the same assumption regarding the uncertainties. The simple fact of the matter is that when the uncertainties in the x-quantities are *not* negligible, deriving equations for the gradient and intercept becomes considerably more difficult. The problem of fitting lines to data when there are uncertainties in both the x- and y-quantities has been recognised for many years but there is still no general agreement about which method for finding m and c is best. Nevertheless, approaches have been developed to deal with the situation of uncertainties in both x- and y- quantities.[8]

To end this chapter on fitting using the method of least squares, it is worth making two observations:

(i) Least-squares fitting is a very powerful data analysis tool but effort is required to ensure that the data substituted into the equations are worth analysing in the first place. Data with large scatter are always difficult to deal with no matter how sophisticated the analysis tool, and it is possible that in some cases the nature of the underlying relationship cannot be established clearly. Therefore, every effort is required to minimise uncertainties at the data-gathering stage and not to rely on the analysis technique 'saving the day'.

(ii) The importance of drawing a graph of x-y data cannot be overstated. It is easy to enter data incorrectly into a calculator or computer package which ultimately produces ridiculous values for the gradient and intercept. These can go unnoticed unless they are compared with approximate values obtained by plotting a graph and estimating the gradient and intercept of a line drawn through the points with a rule. Calculators and computers are excellent when fast or repetitive calculations are required, but cannot match the eye/brain combination when it comes to spotting patterns and anomalies.

PROBLEMS

1. Calculate the gradient and intercept of the x-y data shown in table 6.11 using unweighted least squares. Also calculate the uncertainty in the gradient and intercept.

---

8. A good review of a variety of approaches for finding gradient and intercept when both x- and y-values are subject to uncertainty is 'Least-squares fitting when both variables contain errors: Pitfalls and possibilities' J. R. Macdonald and W. J. Thompson published in the *American Journal of Physics* (1992) volume 60 pages 66–73.

Table 6.11: Linearly related x-y data

x	y
45	23
39	35
31	39
24	47
18	56
11	63
4	75

2. In an experiment to study the optical behaviour of a material, the intensity of light, I, reflected from a crystal of lithium fluoride was measured as the angular position, θ, of a polariser placed between the light source and the crystal was changed. Table 6.12 gives data of the intensity (in arbitrary units) as a function of the angle.

Table 6.12: Variation of intensity of light reflected from a lithium fluoride crystal as a function of the angular position of a polariser

θ (degrees)	Intensity (arbitrary units)
0	1.86
20	1.63
40	1.13
60	0.52
80	0.16
100	0.00
120	0.57
140	1.11
160	1.56
180	1.71

Assume that the function that describes the above data is of the form:

$$I = (I_{max} - I_{min})\cos 2\theta + I_{min}, \text{ where } I_{max} \text{ and } I_{min} \text{ are constants.}$$

(i) What would you plot in order to obtain a straight line for the data shown in table 6.12?

(ii) Fit a straight line to the data and determine the gradient, m, and intercept, c, assuming an unweighted least squares fit is appropriate. Give m and c to three significant figures.

(iii) Calculate the uncertainty in m and c. Give m and c and associated uncertainties to an appropriate number of significant figures.

(iv) Using the gradient and intercept, calculate I_{max} and I_{min}.

3. In a corrosion experiment, the mass of oxide formed on a metal is measured as a function of time that the metal is exposed to air. Table 6.13 gives the data gathered during the experiment.

Table 6.13: Mass of oxide formed as a function of time in a corrosion experiment

time (h)	mass (mg)
0.5	6 ± 1
1	9 ± 1
2	12 ± 1
3	15 ± 1
4	17 ± 1
5	18 ± 1
6	20 ± 1
7	20 ± 1
8	22 ± 1

Assume that the relationship between mass, M, and time, t, can be written:

$$M = \sqrt{kt + D}$$

(where k and D are constants).

(i) Linearise the above equation, indicate what you would plot in order to obtain a straight line, and state clearly how the gradient and intercept of the line are related to k and D.

(ii) Should you use an unweighted or weighted fit when fitting a straight line to these data?

(iii) Using the appropriate weighting, find the gradient and intercept of the best line through the transformed data.

4. The relationship between the mass of gas adsorbed per unit area of surface, g, and the gas pressure, p, can be written (assuming constant temperature):

$$\frac{p}{g} = \frac{1}{n} + \frac{s}{n}p$$

where s and n are constants. Table 6.14 gives experimental data obtained from an adsorption experiment carried out at $0°C$.

Table 6.14: Variation of mass of gas adsorbed per unit area as a function of pressure

g (kg m^{-2})	p (N m^{-2})
1.40×10^{-4}	0.28
1.76×10^{-4}	0.40
2.21×10^{-4}	0.61
2.78×10^{-4}	0.95
3.28×10^{-4}	1.70
3.84×10^{-4}	3.40

Use unweighted least squares to determine the values of n and s. Hint: first rearrange the equation into the form $y = mx + c$.

5. Observations are made of the volume, V, occupied by a particular liquid at temperature, θ, and the values recorded as shown in table 6.15.

Table 6.15: Values of volume occupied by a liquid as the temperature of the liquid increases

V (cm^3)	θ (°C)
1.032	10.0
1.063	20.0
1.094	29.5
1.125	39.5
1.156	50.0
1.186	60.5
1.215	69.5
1.244	79.5
1.273	90.0
1.300	99.0

Assume that $V = 1 + B\theta + D\theta^2$, where B and D are constants.

(i) Linearise the above equation.

(ii) Use the method of least squares to find values for B and D and the uncertainties in these quantities, assuming that the uncertainties in volume values are constant and there is negligible uncertainty in the temperature measurement.

6. The relationship between the current, I, and the voltage, V, of a particular semiconductor device can be written:

$$I = I_0 \exp \left(\frac{eV}{nkT} \right)$$

where:

e is the magnitude of the charge on an electron
(= 1.60×10^{-19}C)
k is Boltzmann's constant (= 1.38×10^{-23} J K^{-1})
T is the temperature in kelvins
I_0 is the saturation current (which is a constant)
n is a constant.

Table 6.16 shows a set of data of current as a function of voltage when the device is maintained at a temperature of 300 K.

Table 6.16: Experimental current and voltage values for a semiconductor device

I (A)	V (V)
7.53×10^{-5}	0.50
3.17×10^{-4}	0.55
1.07×10^{-3}	0.60
3.75×10^{-3}	0.65
1.35×10^{-2}	0.70
4.45×10^{-2}	0.75
1.75×10^{-1}	0.80
5.86×10^{-1}	0.85

(i) Linearise the above equation.
(ii) The uncertainty in each value of current in table 6.16 is equal to 2% of that current. Should a weighted or unweighted least-squares fit to data be performed?
(iii) Perform the appropriate fit to find I_0 and n.

7. The depth of a layer of rock can be determined by generating sound waves at the surface of the Earth. Microphones are spread out over the surface of the Earth to pick up sound waves that have been reflected from layers of rock beneath the Earth's surface.

Table 6.17 shows the distance, d, of each microphone from the point at which the sound was generated and the time delay, t, before the reflected sound wave was detected.

Table 6.17: Time delay of detection of sound wave as a function of microphone position

d (m)	Time delay, *t*, (ms) ± 0.05 ms
15.0	35.36
20.0	36.30
25.0	37.08
30.0	38.22
35.0	39.69
40.0	41.32
45.0	42.90
50.0	44.50

The relationship between *d* and *t* is:

$$t^2 = \frac{d^2}{v^2} + \frac{4h^2}{v^2}$$

where v is the velocity of the wave, and h is the depth of the layer of rock.

(i) What would you plot in order to obtain a straight line?

(ii) Perform a weighted fit to the data to find the gradient and intercept of the graph and the uncertainties in these quantities.

(iii) Using the values determined in part (ii), calculate v and h.

REPORTING EXPERIMENTS

7.1 Overview: purpose of a report

There is often a feeling of accomplishment combined with a sense of satisfaction when an experiment has been completed. The 'hard work' has been done and it is time to relax. However, the reason for the experiment, how it was performed and *what* was discovered may be known to a few people at most. The laboratory notebook containing many of the experimental details is probably only fully comprehensible to its owner and therefore unlikely to provide an easily readable account of the experiment. In situations where all aspects of the experiment need to be communicated to others, a crucial stage of the work remains to be undertaken. That stage is the preparation of a written report.

A report which goes beyond the contents of a laboratory notebook is sometimes referred to as a 'long' or 'formal' report. It is not something that you would produce on completion of every laboratory session you attend. Indeed, in most situations, especially in the early years of study in science and engineering, it is the proper use of a laboratory notebook that is the focus of attention with regard to recording and communicating the work done in the laboratory. Nevertheless, it is quite common to be asked to 'write up' one or more experiments as a formal report, even in the first year of study at university or college.

As you progress through a course of study in science or engineering, more emphasis is likely to be placed upon the communication of experimental results in the form of a report to an instructor or supervisor. At the same time, the report will probably constitute an ever greater proportion of the assessment of any laboratory work you carry out. Beyond this we should recognise that the writing of reports is an everyday activity practised by scientists and engineers; and an effective, thorough and well-written report can be a vital factor when decisions are made about providing support so that the work can continue. It is for these reasons that we will concentrate upon report writing in this chapter.

Writing good reports, like proficiency in performing experiments, requires practice, and first efforts can usually be improved upon. Actively looking for the strengths and weaknesses in reports written by others is a good way to identify what makes a competent report, but is no real substitute for the experience gained through writing one.

A report should:

(i) be complete but concise

(ii) have a logical structure

(iii) be easy to read.

It is possible for a report to be read by people with differing backgrounds and interests. Some readers may have performed experiments similar to those discussed in the report and wish to compare the experimental method and results with their own. For other readers much more basic questions are likely to come to mind:

(i) What was the problem studied and what is its significance?

(ii) What conclusions have been drawn, do they seem reasonable and are they supported by the data?

In addition to the important questions above, many details must be considered in a report, from describing the background to the work to detailing the equipment used. A well-structured report is required to deal with these matters properly. It is this that we will now consider.

7.2 Structure of a report

Report writing requires that all aspects of an experiment be reviewed so that a logical and consistent account of the work can be prepared. It is perhaps not surprising that, at this stage, shortcomings are sometimes identified in the method or analysis, which demand that further work be done before the report can be written.

The layout of the report has a big impact on its clarity and therefore it is usual for a report to be divided into the following sections.[1]

---

1. Where a report is short it is reasonable to combine two or more sections together under one heading. For example, some reports may have a section headed 'Results and Discussion'. Also, some sections may not be applicable such as 'Acknowledgements' and 'Appendixes' and therefore not appear in the report.

    (i) Title
   (ii) Abstract
  (iii) Introduction
  (iv) Materials and methods
   (v) Results
  (vi) Discussion
 (vii) Conclusion
(viii) Acknowledgements
  (ix) Appendices
   (x) References

We will deal with each section in turn, but first we begin by considering the use of English in report writing.

7.3 Use of English

It is unlikely that anyone reads a scientific or technical report specifically to assess the correctness of the language. Nevertheless, if the use of English is poor, a reader may be forced to cover some passages several times in order to find out what the writer is trying to say and, in extreme cases, may give up in frustration.

A number of points can be made which should assist in the task of making a report 'readable'.

Writing in the third person

Though some people choose to write a report in the first person — for example, *I measured the pressure at time intervals of 30 s,* the generally accepted convention is to write in the third person — for example, *the pressure was measured at time intervals of 30 s.* Writing in the third person can leave the reader wondering *who* actually did the things described in the report, especially if there is more than one author. However, this detail is usually of little long-term interest or importance.

Choice of tense

The report is generally written in the past tense when giving details of what was done during the experiment. Occasional use is made of the present tense, especially when giving details of the background to the work or when inferring general relationships from the data. For example:

> *Measurements* **were** *made of the length of the copper rod as a function of temperature. The graph of the data shown in figure 1* **indicates** *that the increase in length of the rod* **is** *directly proportional to the temperature rise.*

Sentence length

Keep sentences short, especially when the content is highly technical or specialised vocabulary is used. The sentence:

The contact resistance was measured by attaching two current leads to the sample through which a current of 1 mA was provided by a constant current source and two voltage leads which were connected to a high input impedance voltmeter as shown in figure 1.

is easier to absorb if it is rewritten as:

To measure the contact resistance, two current leads and two voltage leads were attached to the sample as shown in figure 1. The voltage leads were connected to a high impedance voltmeter. The current leads were connected to a constant current source which provided a current of 1 mA.

Explaining abbreviations fully the first time they are used

It is possible that the analysis technique or instrument used during the experiment is commonly referred to by an abbreviation, such as XRD for X-Ray Diffraction or SEM for Scanning Electron Microscope. Though such abbreviations may be familiar to some readers of the report, to others they will be new and incomprehensible unless fully explained.

Reviewing what has been written

After finishing the first draft of the report, it can be valuable to put it aside for a day or two, if possible, before returning to look it over. This can help reveal omissions and errors that were not obvious when writing the report. Another good way to find out whether the report reads well and 'makes sense' is to ask someone to read and comment on it before it reaches the final version.

7.4 Sections of a report

We will now consider in detail the sections that make up a typical report.

Title

The title of the report should be brief (say between 5 and 15 words) and informative.[2] Consider the following two titles of the

---

2. It is usual to give the name of the author and the date below the title of the report.

same report:

A study of the insulating properties of some materials

and:

A comparison of the thermal insulating properties of styro-foam and fibreglass

The first title is very vague and lacking in information. In the second, we are made aware that it is the *thermal* insulating properties that are to be compared (as opposed to, say, the electrical insulating properties) and specific mention is made of the materials used in the study.

Abstract

This is an overview of the experiment and its findings. It should be brief (50 to 150 words typically) and avoid the detail that a reader will encounter in later sections. The goal is to get straight to the heart of the matter by communicating what was done, why it was significant and what the major findings were. This is not an easy matter and most people prefer to draw up a plan of the whole report, write it, and then return to the abstract later.

Compare the following two abstracts.

Abstract A

The cathodoluminescence of a new ceramic material is discussed in this report. Cathodoluminescence is the emission of light from a material when it is struck with fast moving electrons (like the light emitted from the screen of a television). The light emitted from the new ceramic was analysed, permitting the identification of a compound which is formed at the surface. It is possible to relate the existence of the compound to the difficulties some people have found in making good electrical connections to the ceramic.

Abstract B

Advances in the applications of the superconducting ceramic $YBa_2Cu_3O_{7-\delta}$ have been restricted due to the difficulty of making good electrical connections to this material. In the investigation reported here, cathodoluminescence was used to analyse the surface of the ceramic at points where electrical connections are made. The study offers strong evidence that barium carbonate forms at the surface when the ceramic is exposed to the atmosphere. The barium carbonate forms an insulating layer detrimental to the formation of good electrical connections.

In abstract A, the author has chosen to focus upon the analysis technique used in the investigation (cathodoluminescence) rather than on the main purpose of the experiment which was to find out *why* it is difficult to make good electrical connections to the ceramic material. The abstract goes into the details of the technique: *Cathodoluminescence is the emission of* ... It is important these details be given, but it is preferable to give them in the 'Materials and Methods' section of the report. Finally there is a vague statement to the effect that a compound discovered using the technique can help to explain the electrical connection problem. If the author has uncovered the relationship between the compound and the poor connections, then it should be stated clearly in the abstract.

Abstract B is an improvement upon abstract A. It states:

(i) the reason for the investigation

(ii) the technique used

(iii) what that technique revealed

(iv) how the results obtained can be used to explain *why* electrical connections are so difficult to make to the ceramic material.

Abstract B contains no numerical results and this might be an area in which it could be improved. For example, it might have been worthwhile to include mention of the extent of the layer of barium carbonate.

Introduction

The abstract of the report acts as a summary for the reader. The next stage is to describe the background to the experiment and the particular goals of the experiment or set of experiments. A reader will 'switch off' if too much detail is given and be confused if there is too little. If the work has followed on from someone else's, it is usual to refer to that work. However, the introduction must be understandable to someone not prepared to search out the references. For example, it is not acceptable to say:

> *This work follows on from that carried out by Ms. J. Smith. Please refer to her report for the background and introduction to these experiments.*

The background to the work needs to be clearly described, if applicable, giving an outline of the current understanding of the problem being investigated. The length of the introduction depends on the type of report, but it is unlikely that it would exceed 20% of the whole report. A short report (say 500 words) might have an introduction of about half a page.

An example is given below of an introduction to a short report. Note how it sets the scene, gives some background to the work, and provides a number of references which can be consulted by a reader who wishes to know more.[3]

> *An electrochromic film changes colour when a voltage is applied to the film. When deposited on glass used for windows, such films can be used to control the amount of light entering a room. Consequently, they have great potential for energy efficiency gains when used in industrial or domestic situations. The main methods for depositing the films are evaporation (Durey et al., 1990) and sputtering (Playfair et al., 1990). A third, and possibly more cost-effective method of depositing electrochromic films, is to use the sol-gel technique (Bell et al., 1990). The technique requires inexpensive equipment and can be easily scaled up to provide large area coatings of commercial viability. The technique does have some disadvantages, however. Retention of carbon within films may have a detrimental effect on their optical properties (Bell et al., 1991). This report deals with the deposition of sol-gel films and compares the quality of the films with those prepared using more conventional techniques. The report focuses particularly on the conditions necessary to produce good films and discusses what advances are required before this deposition technique can be made fully viable.*

Materials and methods

This is a description of how the experiment was performed and the materials/samples/components used. All important details, especially diagrams, need to be included in this section of the report. If a standard experimental technique has been used, then it should be described in a few words. Alternatively, a reference should be given where full details of the technique can be found. If the method is new, or a modified version of an established method, then sufficient detail is required to enable someone else to perform the experiment.

Results

It is not necessary to include all the data obtained in the experiment in this section of the report. Doing so may overwhelm the

---

3. How to give references will be dealt with at the end of this section.

reader with table after table of data. Sufficient representative data should be included so that any discussion that follows or conclusions drawn can be seen to be well-supported by the data. Graphs are an excellent way of presenting large quantities of data and are likely to be looked at first, and more carefully, by a reader than data in tabular form. Where calculations are presented, attention must be paid to the impact that uncertainties in the 'raw' data have on the calculated values. The discussion and conclusion will have more credibility if uncertainties have been identified, quantified and accounted for, and their effect on calculated values dealt with explicitly.

Discussion

The discussion section deals with the interpretation of the results that have been presented. An experiment is likely to contain many details, both major and minor. Unless the discussion focuses on the important points, a reader is likely to become lost in a mass of unnecessary detail. Where shortcomings have been identified in the experimental method, these should be discussed. If data from the experiment do not lend strong support to the particular idea or hypothesis at the core of the experiment, then this should be acknowledged. At the very least, the experiment should provide for a better insight into the problem being studied, and other possible means of approaching the problem should have emerged.

Even if the experimental method used could have been improved, we must be careful not to be too dismissive of data that *were* obtained in an experiment. Perhaps better equipment *could* have been used, or *another* week spent collecting data. The question is: what can be usefully said with the data that *were* gathered, despite any shortcomings?

Conclusion

Here we must refer back to the purpose of the experiment. What was the aim of the experiment, and how far did the experiments performed go in achieving that aim? If others have undertaken a similar investigation, then it is usual to include a comparison of findings, giving a reference to the other work. If the value of a quantity has been determined which is sufficiently well known that it appears in a data book or textbook, a comparison of the values should be given along with a reference to the source of the information. For example, as part of a conclusion to an experiment in which the surface tension of a number of liquids was investigated, we might write:

> *An analysis of the data obtained here gives a value of (5.8 ± 0.2) × 10⁻² N m⁻¹ for the surface tension of glycerine at 20°C. This value compares with that published elsewhere of 6.31 × 10⁻² N m⁻¹ (Rowley, 1992).*

Acknowledgements

Experiments in science and engineering are cooperative ventures. There are many situations in which we rely on someone for instruction in the operation of a piece of equipment or the preparation of a sample. Perhaps we have discussed the data with someone in order to clarify our own ideas and this person has assisted in improving our understanding, and hence the report, significantly. Those who have contributed to the work deserve to be acknowledged. If no mention is made of their help, it would not be surprising to find them less than enthusiastic the next time they are called upon to give assistance.

Keep acknowledgements brief, for example:

> *The author gratefully acknowledges the assistance of Ms L. Jones in the area of electronics.*

References

References are an important part of a report. They give the reader access to information concerning the background to the work, details of experimental techniques adopted by others, the results obtained by another experimenter and so on. If the experiment is in an area where there have been many previous publications, it will not be possible to include references to all those publications in the report. Selectivity is the key. If a reference is included to provide general background to the experiment, then a recent book or article should be chosen which offers as wide a view of the subject area as possible. Such a review is likely to have many of its own references which the reader can follow up.

There are two main methods of giving references in a report. The first is to add a superscript number close to the point at which the reference is relevant. For example:

> *The tunnelling of electrons in semiconductors was first reported in the late 1950s.*[1]

In the reference section of the report a consecutive list of numbers appear. Adjacent to each number is the appropriate reference. For example:

[1]Esaki L. 1958 Phys. Rev. **109** 603

The reference is given in a standard form. It includes:
(i) the name(s) of the author(s)
(ii) the year of publication
(iii) the journal or book title (in this example Phys. Rev. is the accepted abbreviation for the journal, Physical Review)
(iv) the volume number, (if it is a journal reference) shown in bold or underlined (in this example, the volume number is 109)
(v) the beginning page number.

It can also be helpful to include the full title of the article in the references. This can lead to the reference section of a report covering many pages if an extensive reference list has been provided.

Another way of referencing in a report is to state the name of the author and the year of publication of the reference at the appropriate point in the text. For example:

A study of the distribution of drugs in the blood and tissues often involves the addition of tracers to the blood followed by measurements of tracer concentration as a function of time (Nichols et al., 1986).

The names are listed in the reference section in alphabetical order followed by the details of date, journal, page number and so on. The advantage of this approach over numbering the references is that if another reference needs to be added to the report, it can be inserted easily without requiring that the remaining references be renumbered.

Appendixes

In a long report some material may need to be included which would affect its readability, such as the derivation of an equation. If no adequate reference can be found, the derivation can be included in an appendix. The listing of a computer program written to assist in the analysis of data is another example of an item which is usually best included in an appendix.

7.5 Comment

Experimental findings in science and engineering are communicated by the means of reports. A report is a permanent record of work that can be referred to and used by others. It could be argued that without proper communication of the purpose, results and findings of an experiment, the effort expended performing the experiment has been largely wasted. When the minute details of an experiment are long forgotten, the report endures as a full account of what was done and is the most visible measure of how well it was done.

PROBLEM

Presented here is a short report of an experiment in which two methods of measuring the acceleration due to gravity are compared. The report is laid out in the manner described in this chapter. Most sections of the report can be improved. Read the report and then concentrate on each section and answer the following questions:

(i) Is it clear?

(ii) Does it contain the relevant information?

(iii) Are there obvious omissions from any section?

(iv) Could anything be omitted without loss?

(v) Could the wording be improved?

Measurement of the acceleration due to gravity: two methods compared

G. Smith

Abstract

An object falling freely under the action of gravity close to the Earth moves with constant acceleration. This report assesses two methods for determining the acceleration due to gravity, g. One approach uses the relationship between g and the time for an object to fall a fixed distance. The other is based on the measurement of the period of a simple pendulum. Both methods yield values for g which are consistent with those reported elsewhere, but the method adopted using the simple pendulum is shown to be more accurate and precise.

Introduction

All bodies on the surface of the Earth experience a force due to gravity. The effect of the force is most obvious when a body is allowed to fall freely under the influence of gravity. Newton's law of gravitation predicts that the acceleration due to gravity, g, is dependent on the mass of the Earth and the distance that the falling body is from the centre of the Earth. As a body close to the Earth's surface falls a small distance compared to the radius of the Earth, g can be regarded as constant over that distance. The aim of this experiment is to compare two methods for establishing the value of g near to the surface of the Earth.

Materials and methods

Method A

The time for a small metal ball to fall through various distances, h, was measured. Figure 1 shows the arrangement used.

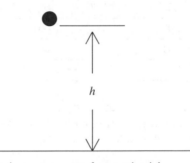

Figure 1: Experimental arrangement for method A

The distance h was measured with a tape measure which has a smallest division of 1 mm. The timing of the fall was done by hand using a stopwatch with a resolution of 0.01 s. In order to obtain a better estimate of the time of fall, five repeat measurements of the time of fall were made at each distance.

Method B

A small ball was attached to a fixed point by a thread. The ball was then allowed to swing through a small angle (<5°) as shown in figure 2.

Figure 2: Experimental arrangement for method B

143

A stopwatch was used to measure the period of the motion, T, as a function of the length, L, of the pendulum. To reduce the influence of reaction time, the time for five successive swings was measured.

Results

Method A

Table 1 shows the data gathered for the free fall of the ball as a function of distance. The second column of the table contains the mean of the five measurements made at each distance, h. The uncertainty in the time was calculated by dividing the range of times measured at each distance by the number of repeat measurements (5).

Table 1: Time for ball to fall various distances

Distance, h (m) ± 0.01 m	Time (s)	Uncertainty in time
0.86	0.38	±0.06
2.02	0.65	±0.06
3.01	0.79	± 0.1
4.26	0.99	± 0.1

The relationship between distance, h, through which a body moves (beginning from rest) with constant acceleration, a, in time, t, is given by (Resnick et al., 1992):

$$h = \tfrac{1}{2}at^2 \qquad (1)$$

This can be rearranged to give:

$$t = \left(\frac{2}{a}\right)^{\frac{1}{2}} h^{\frac{1}{2}} \qquad (2)$$

This equation is of the form $y = mx + c$, with intercept, $c = 0$ and gradient, m, given by:

$$m = \left(\frac{2}{a}\right)^{\frac{1}{2}} \qquad (3)$$

In this experiment the acceleration in (3) is equal to g.

Figure 3 shows a plot of t versus $h^{\frac{1}{2}}$ for the data in table 1.

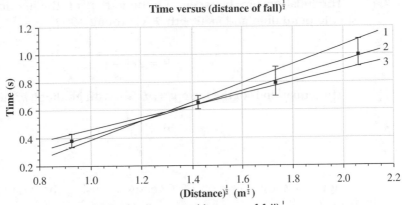

Figure 3: Time of fall of ball versus (distance of fall)$^{\frac{1}{2}}$

Table 2 shows the gradient of the three lines drawn through the data points along with the corresponding values for g based on equation 3.

Table 2: Gradient of lines shown in figure 3

Line	gradient m $(\text{s m}^{-\frac{1}{2}})$	acceleration due to gravity, g (m s^{-2})
1	0.64	4.9
2	0.51	7.7
3	0.40	12.5

The acceleration due to gravity using this method is found to be $(8 \pm 4)\ \text{m s}^{-2}$.

Method B

Table 3 shows the data gathered for the period of the pendulum as a function of the length of the pendulum.

Table 3: Measured values of period of pendulum as a function of pendulum length

Length (m) $\pm$ 0.005 m	Period (s) $\pm$ 0.02 s
0.410	1.32
0.575	1.55
0.700	1.71
0.825	1.84
0.900	1.93

The relationship between the period, T, of the motion of a simple pendulum and its length, L, is (Young 1992):

$$T = 2\pi\left(\frac{L}{g}\right)^{\frac{1}{2}} \qquad (4)$$

This equation is of the form $y = mx + c$ with gradient, m, given by:

$$m = \frac{2\pi}{g^{\frac{1}{2}}} \qquad (5)$$

Figure 4 shows a graph of T versus $L^{\frac{1}{2}}$ using the data given in table 3.

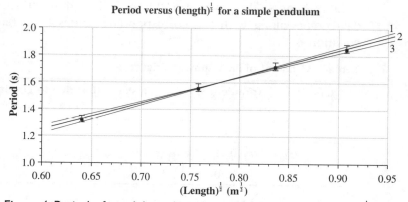

Figure 4: Period of pendulum plotted against (length of pendulum)$^{\frac{1}{2}}$

Table 4 shows the gradient of the three lines drawn through the data points along with the corresponding values for the acceleration due to gravity based on equation 5.

Table 4: Gradients of lines shown in figure 4

Line	gradient $(\text{s m}^{-\frac{1}{2}})$	acceleration due to gravity (m s^{-2})
1	2.12	8.78
2	2.00	9.87
3	1.88	11.2

The acceleration due to gravity using this method is found to be (9.9 ± 1.2) m s^{-2}.

Discussion

Values for the acceleration due to gravity, determined using both methods, are consistent with those published elsewhere (Rocke 1984). However, the method using the pendulum is superior in so far as the 'best' value obtained is close to those reported by other workers and the uncertainties in the value of g are less than that of the first method, that is method B is both more precise and more accurate than method A. In both methods the main source of uncertainty is in the timing of the events. However, in the case of the pendulum, the reaction time was a small fraction of the total time measured, whereas in method A the reaction time was a large fraction of each timing. An electromechanical or optical means of synchronising the beginning and end of the motion ball with the timing instrument is required if method A is to provide a more accurate value for g.

Conclusion

A value of (9.9 ± 1.2) m s^{-2} was determined for the acceleration due to gravity by study of the motion of a simple pendulum. This compares with a value of (8 ± 4) m s^{-2} found by measuring the time for an object to fall freely under gravity. The pendulum method is preferred due to its superior accuracy and precision.

References

Resnick R., Halliday D. And Krane K. S. (1992) *Physics* 4th ed. (Wiley: New York) p. 55
Rocke F. A. (1984) *Handbook of Units and Quantities* (Atomic Energy Commission: New South Wales) p.110
Young H. D. (1992) *University Physics* 8th ed. (Addison Wesley: Massachusetts)

USING A POCKET CALCULATOR FOR DATA ANALYSIS

8.1 Overview: the role of the pocket calculator in data analysis

Pocket calculators have an important role to play in the analysis of experimental data. Over the past 20 years pocket calculators have become so commonplace that it is unusual to find anyone in a scientific or engineering discipline who does not own one. At the same time the number of functions that each can perform has increased dramatically. Even an inexpensive scientific calculator offers a wide range of useful mathematical and statistical functions. A price must be paid for the increased scope of the calculator: what *are* the less well-known functions and how do you use them? The problem of how to use the uncommon functions is further complicated by the calculator manuals. Many have only a brief description of each function and are written in five or more languages. A significant problem is that the manuals are so small that they are easily misplaced!

Despite the drawbacks, pocket calculators can assist greatly in the analysis of experimental data, and possess the advantages of reliability, compactness and portability. In this chapter we will discuss how to use a calculator to find means and standard deviations, and how to use one to perform linear regression.

One of the biggest manufacturers of calculators is CASIO and the following discussion is based on the data analysis features of the scientific calculators manufactured by this company. If you have a Sharp, Texas, Hewlett Packard or other type of (scientific) calculator you will normally find that they possess many of the functions that we discuss in this chapter, although they may be labelled differently. You will need the instruction manual on hand so that you can check out how to use these functions.

Operational details in the forthcoming sections are specifically relevant to the CASIO *fx-100* series of calculators (for example, *fx-100D, fx-100V*). These models are very popular with students, although other models such as the *fx-82* and *fx-350* have many features in common with the *fx-100*. You may have to search the keyboard above **and** below the keys on the keypad to find the function you are looking for. If you do not have a calculator but are considering buying one, then I can recommend the CASIO *fx-100D* as being very good for data analysis and for many other applications in science and engineering.

8.2 Range and rounding errors

All calculators (and computers) are limited in some way as to the range of numbers they can cope with. For example, many pocket calculators would be unable to calculate 5×10^{45} multiplied by 6×10^{61}. This is because when the number resulting from a calculation is equal to, or greater than, 1×10^{100}, the upper limit of the calculator is exceeded and an error (denoted by *-E-*, or something similar) appears on the display. In practice, this is a minor restriction, and the range of numbers that a pocket calculator can deal with is more than adequate for most purposes in experimental science and engineering.

The number of digits that a calculator holds internally when it represents a number (its internal precision) may have an effect on some calculations, especially in situations where calculations involve the subtraction of two numbers that are nearly equal. As an example, consider the following calculation. Suppose m is given by:

$$m = \frac{331.276 - 41.235}{8017.65 - 8016.55} = 263.674$$

If the calculator rounds the numbers used in the division to four figures (which, happily, they normally don't!), we would find that $m = 290.1$, which differs considerably from 263.674. Check this out by rounding the numbers above to four significant figures (so that, for example, 331.276, in the above calculation, becomes 331.3).

Most modern pocket calculators can display up to ten digits at any time and keep two 'in reserve' to ensure that the majority of calculations are correct up to ten figures. As it is rarely justified in experimental situations to quote 'final answers' to more than five significant figures, calculator rounding errors do not have an effect.

When many repetitive calculations are carried out *automatically* by pocket calculators (usually those that are 'programmable') — say when an iterative process is being used to find the solution to an equation — rounding errors can become important. However, it is fair to say that the calculations you are likely to attempt, based on work covered in this book, are unlikely to be adversely affected by the precision capabilities of your calculator.

8.3 Switching modes

Calculator manufacturers are faced with the problem of how to offer more functions on their calculators while still making them relatively easy to use. The manner by which you switch to a different *group* of functions on a CASIO calculator is to use the MODE key, followed immediately by pressing a number. For example, on the *fx-100*, pressing MODE followed by 5, switches the calculator into radian mode. Thereafter, the use of any trigonometrical function, such as sine, assumes that the number entered is an angle in radians.

8.4 The SHIFT key

In order to provide more options on their calculators, manufacturers are often forced to use each key on the calculator to serve more than one purpose. The consequence is that the keypad looks cluttered with numbers, letters and symbols. The function appearing above a key becomes active after pressing the SHIFT key. CASIO calculators indicate that the SHIFT has been activated by showing S in the top left-hand corner of the display.

As an example of using the SHIFT key, let us calculate e^{22}. The key e^x is found above the ln key and to find e^{22}, we use the following key sequence:

Press 22 SHIFT ln   The display should show 3584912846

After switching the calculator into statistics mode, we will need to use the SHIFT key to access some of the statistical functions such as the mean and standard deviation.

8.5 Data analysis functions

Consider the data set shown in table 8.1 which gives the temperatures indicated by 12 thermometers that have been placed close together in a well-stirred bath of water.

Table 8.1: Twelve values of the temperature of a water bath

Temp. (°C)	15.2 15.1 15.0 14.8 15.1 15.2 15.1 15.0 14.6 15.5 15.3 15.2

We could calculate the mean and standard deviation of the numbers above in a step-by-step manner using the formulae we discussed in chapters 4 and 5, but as the calculator has these functions *built-in*, it makes sense to use these.

After switching the calculator on, press SHIFT AC to clear the memory of the calculator.[1] This is done because they have the annoying habit (which can sometimes be useful) of remembering numbers that have been previously stored in memory, even after the calculator has been switched off. Pressing SHIFT AC clears the memory of all numbers, ready for new ones to be entered. To switch to the data analysis mode, press MODE 3 and you should see the letters SD appear along the top row of the display of the calculator. To enter the numbers press the following keys:

15.2 M+ 15.1 M+ 15.0 M+ 14.8 M+ 15.1 M+ 15.2 M+
15.1 M+ 15.0 M+ 14.6 M+ 15.5 M+ 15.3 M+ 15.2 M+

You need to be careful when entering numbers because there is no way to check the numbers *after* they have been entered, so a mistake can easily go undetected. If you *do* have to delete a number there are two approaches:

(i) If you have keyed in the wrong number but have not entered it by pressing the M+ key, simply clear the display by pressing C and key in the correct number followed by pressing the M+ key.

(ii) If you have already entered the number, you need to delete that number from the memory of the calculator by pressing SHIFT M+. You should remove the offending number immediately otherwise you will lose track of the values you *have* entered.

Note that here lies a disadvantage of many calculators. After entering the data, it is not easy to check whether you have made a mistake. As the process of entering data is not sensitive to the order in which you enter the numbers, a good tip is to clear the memory of the calculator and re-enter the numbers in reverse order. If the numbers you obtain *from* the calculator (for example the mean) are

---

1. There is an 'independent memory' which cannot be cleared by pressing SHIFT AC. However, the memory used for statistical calculations *is* cleared in this manner.

the same both times then it is probably safe to assume that no mistake has been made when keying in the numbers.

You can check *how many* numbers you have entered by using the $\boxed{\text{KOUT}}$ key. $\boxed{\text{KOUT}}$ gives you access to various useful stored quantities such as the sum of the numbers entered (Σx) and the sum of the *square* of the numbers (Σx^2). To find out how many numbers you have entered:

Press $\boxed{\text{KOUT}}$ 3   The display should show 12

The mean of the 12 numbers you have entered is found by using the $\bar{x}$ function which requires that you press the $\boxed{\text{SHIFT}}$ key. On the *fx-100*, $\bar{x}$ is found above the 1 key so:

Press $\boxed{\text{SHIFT}}$ 1   The display should show 15.09166667

In most of the examples given in this chapter all the figures that appear on the calculator display are presented (so that you can compare the numbers appearing on your display with those given here). However, in most situations you will need to *round* the number shown to a sensible number of figures, as discussed in chapter 2.

EXERCISE A

Switch your calculator into statistics mode ($\boxed{\text{MODE}}$ 3 on the CASIO *fx-100*) and enter the following numbers: 42, 43, 51, 76, 48, 91, 31, 92. Use the built-in function on your calculator to determine $\bar{x}$.

8.5.1 Standard deviation

CASIO calculators possess two standard deviation functions, labelled $x\sigma_{n-1}$ and $x\sigma_n$ or, in some cases, σ_{n-1} and σ_n. We have discussed these functions previously in chapter 5 where they were represented by the symbols s (the estimate of the population standard deviation) and σ (the sample standard deviation) respectively. The x before the σ_{n-1} refers to the fact that the data set is represented by the symbol x, and so $x\sigma_{n-1}$ does *not* represent x multiplied by σ_{n-1}. The definitions of $x\sigma_{n-1}$ and $x\sigma_n$ are as follows:

$$x\sigma_{n-1} = \left(\frac{\sum (x_i - \bar{x})^2}{n-1} \right)^{\frac{1}{2}} \text{ (which is the same as } s \text{ in chapter 5)}$$

$$x\sigma_n = \left(\frac{\sum (x_i - \bar{x})^2}{n} \right)^{\frac{1}{2}} \text{ (which is the same as } \sigma \text{ in chapter 5)}$$

Example

The following numbers were entered into the calculator in the manner described in section 8.5 (remember to press $\boxed{\text{SHIFT}}$ $\boxed{\text{AC}}$ before entering the numbers):

$$1.2, 3.3, 1.4, 1.2, 1.8, 1.9, 1.6, 2.1, 1.5$$

To calculate $x\sigma_n$:

press $\boxed{\text{SHIFT}}$ 2. The display should show 0.610605851

To calculate $x\sigma_{n-1}$ for the data we have already entered:

Press $\boxed{\text{SHIFT}}$ 3   The display should show 0.647645307

EXERCISE B

Use the built-in functions of your calculator to find σ ($x\sigma_n$) and s ($x\sigma_{n-1}$) for the following data. Give the values of σ and s to the full precision of your calculator and the same values rounded to *one* significant figure.

$$0.034, 0.056, 0.023, 0.043, 0.087, 0.101, 0.009, 0.086$$

8.5.2 Other useful quantities

The calculator gives us access to other useful quantities such as Σx and Σx^2. These are especially useful when doing linear regression because they are required when we calculate the uncertainties in the gradient and intercept of a straight line. To output Σx and Σx^2 we need to use the $\boxed{\text{KOUT}}$ button on the calculator. To find Σx for the data shown in table 8.1:

Press $\boxed{\text{KOUT}}$ 2   The display should show 181.1

To find Σx^2:

Press $\boxed{\text{KOUT}}$ 1   The display should show 2733.69

EXERCISE C

With your calculator in statistics mode enter the following numbers: 251, 345, 720, 120, 560, 500, 403. Use the built-in functions to find Σx and Σx^2.

8.6 Least-squares fitting

In chapter 6 we discussed fitting a straight line to a set of *x-y* data. The calculations are cumbersome to do 'by hand' and it is easy to make an arithmetical error. The facility for unweighted least-squares fitting is now commonly found on scientific calculators. The CASIO *fx-100* series allows you to fit a straight line to data and to obtain the values of gradient of the line and intercept on the *y*-axis.

8.6.1 Linear function

In chapter 3 we found that the linear relation between the dependent variable, *y*, and independent variable, *x*, may be written in the form:

$$y = mx + c$$

where *m* is the gradient of the line and *c* is where the line inter-cepts the *y*-axis (i.e. the value of *y* where $x = 0$).

CASIO and other calculator manufacturers tend to use different symbols to represent gradient and intercept. CASIO write the above equation in the form:

$$y = Bx + A, \qquad \text{so that } B \equiv m \text{ and } A \equiv c$$

The formulae that the calculator uses to calculate *B* and *A* are those given by equations 6.7 and 6.8 respectively in chapter 6. Although uncertainties in *B* and *A* cannot be found directly using the calculator, the information needed to calculate these quantities is available in the calculator memory.

To switch to least-squares mode, press $\boxed{\text{MODE}}$ 2. The letters LR (standing for linear regression) should appear along the top row of the display. Before beginning to enter any data, clear the memory by pressing $\boxed{\text{SHIFT}}$ $\boxed{\text{AC}}$.

Let us now consider an example of using the least-squares fitting functions on the calculator. In table 8.2 we give some data that were gathered during a study of the elasticity of a material. Increasing loads were added to a wire made from the material and the total length of the wire was recorded after each increase in load.

Table 8.2: Total length of a wire under various loads

y (length of wire in cm)	16.1	16.5	17.2	17.7	18.1	18.9
x (load in g)	100	120	140	160	180	200

As it is the length of the wire that increases *as a result* of applying a greater load, the load is the independent (*x*) variable and the length of the wire is the dependent (*y*) variable.

To enter the data into the calculator, first type the *x*-value followed by pressing the $\boxed{\text{[(---]}}$ button, then type the corresponding *y*-value followed by pressing the $\boxed{\text{M+}}$ key. For the data above we enter the data as follows (Note: it is important to be extra careful when entering data as mistakes are usually difficult to detect.):

100	$\boxed{\text{[(---]}}$	16.1	$\boxed{\text{M+}}$
120	$\boxed{\text{[(---]}}$	16.5	$\boxed{\text{M+}}$
140	$\boxed{\text{[(---]}}$	17.2	$\boxed{\text{M+}}$
160	$\boxed{\text{[(---]}}$	17.7	$\boxed{\text{M+}}$
180	$\boxed{\text{[(---]}}$	18.1	$\boxed{\text{M+}}$
200	$\boxed{\text{[(---]}}$	18.9	$\boxed{\text{M+}}$

Now all the data are entered. The function which gives you the gradient of the best straight line fitted to these data is labelled *B* and it can be found above the 8 key. To find the gradient, *B*:

Press $\boxed{\text{SHIFT}}$ 8   The display should show 0.027571428

so that the gradient is 0.027 571 428 cm g^{-1}.

To find the intercept of this line on the *y*-axis, *A*:

Press $\boxed{\text{SHIFT}}$ 7   The display should show 13.28095238

so that the intercept is 13.280 952 38 cm.

EXERCISE D

The voltage across a resistor is measured as the current through it is increased. Table 8.3 shows the current and voltage data.

Table 8.3: Current and voltage data for a resistor

Current (mA)	5.5	10.8	14.7	19.7	25.1	29.8	32.6	40.1
Voltage (V)	1.15	2.08	2.98	4.10	4.88	6.06	6.32	7.85

Taking the current as the *x*-quantity and the voltage as the *y*-quantity, and assuming a linear relationship between *x* and *y*, calculate the gradient (*B*) and the intercept (*A*). Give *B* and *A* to the full precision of your calculator and then round these values to three significant figures. Don't forget to state the units of measurement in your answer.

8.6.2 Expected values for *x* and *y*

When the best straight line has been fitted to the data, we can find a value for y which corresponds to any given value of x based on the equation of the line. This is very useful if we wish to interpolate between data points (interpolation is discussed in section 3.3.3).

For example, the line that fits the data shown in table 8.2, is:

$$y = 0.027\ 57x + 13.28$$

where we have rounded values of A and B to four significant figures. To know how many figures to round our values of A and B to, we must find the uncertainties in these values. We will cover this in the next section.

To find the value of y when $x = 135$ g, for example, we can simply substitute 135 for x into the equation given above. However, the value of y can more quickly be calculated using the built-in function on the calculator. As an example, enter the *x-y* data shown in table 8.2. Next:

Press 135 $\boxed{\cdots)}$ The display should show 17.00309524

This means that when $x = 135$ g, $y = 17.003\ 095\ 24$ cm.

If we want the value of x corresponding to a particular value of y we need to use the $\boxed{\text{SHIFT}}$ key as follows. Suppose we wish to know the value of x when $y = 18$ cm, we would:

Press 18 $\boxed{\text{SHIFT}}$ $\boxed{\cdots)}$ The display should show 171.1571676

This means that when $y = 18$ cm, $x = 171.157\ 167\ 6$ g.

8.6.3 Uncertainties in gradient and intercept

In section 8.6.1 we considered an example in which the gradient, B, turned out to be equal to $0.027\ 571\ 428$ cm g^{-1} and the intercept, A, was equal to $13.280\ 952\ 38$ cm. If we wish our work to be taken seriously we know we cannot quote A or B to this many significant figures. In section 6.2.3 we discussed two equations which allow us to calculate the uncertainty (the standard error) in A and B, σ_A and σ_B. We cannot find these values in 'one step' using a calculator but it is possible to calculate the uncertainties without too much difficulty. First, let us remind ourselves of the formulae needed to calculate σ_A and σ_B, where these are the uncertainty in the intercept and gradient respectively. We have that:

$$\sigma_B = \frac{\sigma n^{\frac{1}{2}}}{[n\sum x^2 - (\sum x)^2]^{\frac{1}{2}}} \tag{8.1}$$

and

$$\sigma_A = \frac{\sigma(\sum x^2)^{\frac{1}{2}}}{[n\sum x^2 - (\sum x)^2]^{\frac{1}{2}}} \tag{8.2}$$

where n is the number of data points.

We need to calculate σ^2. First we subtract the calculated values for y, y_{ic}, for each value of x, from the observed values of y, y_{io}, as shown in table 8.4. This gives us the residuals which we square and sum.

Table 8.4: Columns required for calculation of σ

x (g)	y_{io} (cm)	y_{ic} (cm)	$y_{io} - y_{ic}$ (cm)	$(y_{io} - y_{ic})^2$ (cm^2)
100	16.1	16.038 10	0.061 905	0.003 832
120	16.5	16.589 52	−0.089 520	0.008 015
140	17.2	17.140 95	0.059 048	0.003 487
160	17.7	17.692 38	0.007 619	0.000 058
180	18.1	18.243 81	−0.143 810	0.020 681
200	18.9	18.795 24	0.104 762	0.010 975

The sum of the square of the residuals is 0.047 048. To calculate σ^2, we use the relationship we met in chapter 6, namely,

$$\sigma^2 = \frac{\sum (y_{io} - y_{ic})^2}{n - 2}$$

It follows that, $\sigma^2 = \dfrac{0.047\ 048\ \text{cm}^2}{4} = 0.011\ 87\ \text{cm}^2$

With the CASIO *fx-100* calculators, the numbers generated by least-squares fitting occupy all but one of the memories available in the calculator, so we need to record Σx and Σx^2 before switching back to the normal mode of the calculator to do the remaining calculations. After entering the data given in table 8.2, Σx^2 is found in the following way:

Press $\boxed{\text{KOUT}}$ 1   The display should show 142000

To find Σx:

Press $\boxed{\text{KOUT}}$ 2   The display should show 900

Calculating $n\sum x^2 - \left(\sum x\right)^2$ gives:

$$6 \times 142\ 000\ \text{g}^2 - (900\ \text{g})^2 = 42\ 000\ \text{g}^2$$

Now we use equation 8.1 to give:

$$\sigma_B = \left(\frac{6 \times 0.011\ 87\ \text{cm}^2}{42\ 000\ \text{g}^2}\right)^{\frac{1}{2}} = 1.302 \times 10^{-3}\ \text{cm}\ \text{g}^{-1}$$

and equation 8.2 to give:

$$\sigma_A = \left(\frac{0.011\ 87\ \text{cm}^2 \times 142\ 000\ \text{g}^2}{42\ 000\ \text{g}^2}\right)^{\frac{1}{2}} = 0.20\ \text{cm}$$

Finally, we can quote gradient and intercept along with the uncertainty in both as:

$B = (2.76 \pm 0.13) \times 10^{-2}\ \text{cm}\ \text{g}^{-1}$ and the intercept as

$A = (13.3 \pm 0.2)\ \text{cm}$.

EXERCISE E

The temperature variation along a tile used for insulation purposes in a furnace is measured in an experiment, and data obtained are shown below.

Table 8.5

Temperature (°C)	120	245	345	476	589	680	790	900
Distance (cm)	1.1	2.3	3.0	4.2	5.5	6.5	7.2	8.5

1. Taking the distance as the x-quantity, and the temperature as the y, use the least-squares facility of your calculator to find the following quantities (quote values to the full precision of your calculator and don't forget to include the units of measurement):
   - (i) the gradient, B
   - (ii) the intercept, A
   - (iii) $\Sigma x, \Sigma x^2$ and Δ
   - (iv) σ^2
   - (v) the uncertainty in B and A.
2. Quote B and A and their uncertainties to an appropriate number of significant figures.

8.7 Limitations of calculators

Calculators are excellent in situations where the number of data values you have to deal with is small (say under 20). As the number of values that must be entered increases, so does the likelihood of making a mistake or losing track of where you have reached. Some calculators store every number in a separate memory location. This permits you to step through the memory locations so that you can look at the numbers and correct them if necessary. This facility is normally only found on fairly expensive calculators. (The CASIO *fx-100* series of calculators does not allow you to do this.) With most calculators you need to identify and correct the mistake before it is entered.

Perhaps the biggest drawback with pocket calculators is their inability to give a histogram or *x-y* graph of entered data.[2] If you are dealing with *x-y* data, a graph of the data reveals immediately whether there are any obvious outliers. You can then investigate whether an outlier is a point that has been plotted incorrectly, or is because of some other reason.

---

2. Some fairly expensive calculators do have a large liquid crystal display that permits the graphical display of data, such as the CASIO *fx-7700GB*.

SPREADSHEETS IN DATA ANALYSIS

9.1 Overview: what is a spreadsheet and how can it be used for data analysis?

The data analysis techniques that we have discussed so far, for example fitting a straight line to x-y data, have required calculations that have been fairly easy to perform using 'pencil and paper' or a pocket calculator. As long as there are not too many data values, and we are careful to avoid arithmetic mistakes in calculations as we proceed, these methods are fine. As the amount of data we have to deal with increases, so does the likelihood that we will make an arithmetic error or enter an incorrect number into our calculator. Some calculations are cumbersome, such as when a weighted least-squares fit to data is done (see section 6.3). Under normal circumstances this requires a table to be drawn up that contains many columns, and demands tens, if not hundreds, of time-consuming calculations.

It would be very convenient to have an 'analysis system' into which we could enter raw data, and perform calculations using that data, such as a least-squares fit, automatically. It would be valuable to be able to show the results of calculations in graphical as well as numerical form, and to allow data to be easily added to and modified. In situations where a computer has been used to assist data gathering, it would be attractive to be able to transfer that data to the system for analysis and presentation purposes, without having to go through the chore of first recording numbers on paper and then entering them 'by hand' into an analysis package.

Systems that have these capabilities *are* available in the form of software packages for computers called *spreadsheets*. Spreadsheets have been available since the early 1980s for microcomputers, and since that time they have developed to an extent where they have become very powerful tools for analysing and presenting all forms of scientific and technical data. In this chapter we will discuss basic features of spreadsheets and show how they may be applied to the analysis of experimental data. At the time of writing there are many spreadsheet packages available for microcomputers, of which three of the most popular are Lotus 1-2-3, Quattro Pro and Excel. Although the advanced features of these (and other) packages differ, they all share the same basic features. We have chosen to focus on Excel (version 5) when specific details are given,[1] though other professional spreadsheet packages would produce similar results.

9.1.1 Illustration of experimental data analysis using a spreadsheet

Figure 9.1 indicates what can be done with a spreadsheet in just a few minutes. An experiment has been performed to study the linearity of a position-sensitive detector. The output voltage from the detector has been recorded as a function of the displacement of a focused light spot on the active area of the detector. Data are entered into the spreadsheet in columns and the data points are plotted in the form of an x-y graph. The spreadsheet has calculated the gradient and intercept of the best straight line using equations 6.7 and 6.8 respectively, and has plotted this line on the graph.

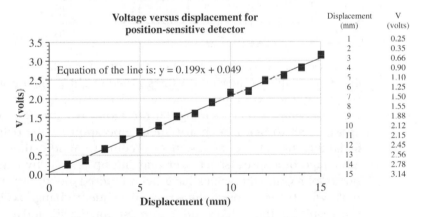

Displacement (mm)	V (volts)
1	0.25
2	0.35
3	0.66
4	0.90
5	1.10
6	1.25
7	1.50
8	1.55
9	1.88
10	2.12
11	2.15
12	2.45
13	2.56
14	2.78
15	3.14

Equation of the line is: y = 0.199x + 0.049

Figure 9.1: Example of graph produced by a spreadsheet

Whenever the data in a spreadsheet are changed (which could be required if a mistake has been made when entering the data), the graph is updated and any computations linked to that data,

---

1. Excel 5 is a spreadsheet developed and distributed by the Microsoft Corporation of America.

such as finding the gradient and intercept of a set of *x-y* data, are also immediately recalculated.

Spreadsheet packages have extensive graph-plotting facilities which is one of their main attractions. With a press of a key, the graph shown in figure 9.1 can be transformed into a log-linear or log-log plot of data. Scaling of the graph is done automatically, with the option to manually override this if you wish.

9.2 Spreadsheet basics

A spreadsheet consists of a two-dimensional array of boxes, often referred to as *cells* as shown in screen 9.1.[2] (To indicate clearly that we are dealing with a spreadsheet, we will present it as a 'screen' and show it bordered by two lines.)

Each cell in the spreadsheet can be identified by a combination of the column letter and the row number.[3] So, for example, we have entered the number 14.751 into the cell **B2**.

Screen 9.1: Small section of a spreadsheet

	A	B
1		
2		14.751
3		

Each cell can contain a number, text or a formula. Thus we can draw up a table of measured values as we might do in a notebook by giving each column a heading and entering the data in the column beneath that heading. Furthermore, we can instruct the spreadsheet to perform an arithmetic operation on the data, such as adding up all the values in one column. Where then, is the advantage of a spreadsheet over drawing up a table 'by hand' in a notebook? Once the values have been entered along with the formula which uses those values, any subsequent change will cause the result of the calculation to be automatically updated. This apparently simple feature is extremely useful when many interlinked calculations need to be performed on a data set.

---

2. Usually, about 10 columns by 15 rows of cells can be displayed on the screen of the computer at any one time, but for convenience we have chosen to show only a small part of the screen in screen 9.1 and the following screens.
3. Some spreadsheets allow you to change how a cell is identified. For example, you could designate columns by numbers instead of letters.

In order to illustrate this, screen 9.2(A) shows a spreadsheet containing the values of ten precision resistors of nominal value 33.0 Ω which have been measured with an instrument of resolution 0.1 Ω.

Screen 9.2(A): Section of spreadsheet containing an incorrectly entered number

	A	B
1		Resistance (Ω)
2		33.4
3		33.8
4		32.2
5		31.9
6		32.2
7		33.9
8		34.0
9		31.4
10		31.6
11		32.3
12	sum =	326 7

Screen 9.2(B): Updated spreadsheet after correcting number in cell B7

	A	B
1		Resistance (Ω)
2		33.4
3		33.8
4		32.2
5		31.9
6		32.2
7		29.3
8		34.0
9		31.4
10		31.6
11		32.3
12	sum =	322.1

The number in cell **B7** (33.9) in screen 9.2(A) has been entered incorrectly and should be replaced by 29.3. Cell **B12** shows the sum of the ten numbers; as soon as the correct number has been entered, the sum is immediately updated from 326.7 to 322.1 as seen in screen 9.2(B).

9.2.1 Calculations involving columns of numbers

Once the raw data from an experiment have been recorded, it is likely that we will want to perform a variety of arithmetic operations on them, such as addition, subtraction, multiplication. Or perhaps we would like the natural logarithm of each recorded value to be calculated. A spreadsheet can do all of this quickly and efficiently.

As an example, consider an experiment in which a detection system has been used to count the number of gamma rays that have penetrated a material. Screen 9.3 shows the number of counts recorded (over a period of one minute) as the thickness of the absorbing material is varied. One step in the analysis to determine the capability of the material to absorb the radiation requires we take the natural logarithm of the recorded counts. This is a tedious business with a calculator, especially when there are many data values. By contrast, this operation is very easy to do on all of the data 'simultaneously' using a spreadsheet.

The first step that is required is to instruct the spreadsheet that cell **C2** should be set equal to the (natural) logarithm of the contents of cell **B2**. In the Excel spreadsheet this is done by making **C2** the active cell[4] and typing into it the formula that will calculate the logarithm of **B2**. To indicate that it is a formula that you are about to enter (and not data or text) you begin with an equals sign, and then follow this immediately with the formula as shown in screen 9.3.[5] Note that in a cell containing a formula, Excel uses uppercase letters to indicate any mathematical function in that cell — for example LN(**B2**), rather than ln(**B2**).

Screen 9.3: Number of counts recorded over one minute for various thicknesses of absorbing material

	A	B	C
	Thickness (mm)	Counts	ln(Counts)
1			
2	1	1213	= LN(B2)
3	2	785	

---

4. To make a cell 'active' you can use the cursor controls on the keyboard of the microcomputer to highlight the cell.

5. This is true for Excel; other spreadsheets use similar methods for entering formulae.

As soon as the enter key is pressed, cell **C2** is replaced by the calculated value, as shown in screen 9.4.[6]

Screen 9.4: Automatic calculation of the natural logarithm of the number of counts shown in cell C2

	A	B	C
1	Thickness (mm)	Counts	ln(Counts)
2	1	1213	7.100851909
3	2	981	

The next step is to use the facility that all spreadsheets possess for filling a group of cells with a selected formula. This allows the logarithms of the remaining numbers to be calculated quickly. In this example a **Fill** command is used to fill **C2** through to **C11** with LN(**B2**) through to LN(**B11**). In Excel this is called the **Fill Down** command. Within a fraction of a second all the logarithms are calculated and appear in the C column as shown in screen 9.5.

Screen 9.5: Use of fill down command in C column

	A	B	C
1	Thickness (mm)	Counts	ln(Counts)
2	1	1213	7.100851909
3	2	981	6.88857246
4	3	752	6.622736324
5	4	631	6.447305863
6	5	423	6.047372179
7	6	491	6.196444128
8	7	329	5.796057751
9	8	259	5.556828062
10	9	215	5.370638028
11	10	158	5.062595033
12	11	141	4.94875989
13	12	118	4.770684624

6. The number that appears in cell C2 is given to 10 figures (the same as for most pocket calculators). In fact, Excel 5 holds numbers internally to 15 figures. These extra figures are very useful for reducing the effect of rounding errors. However, we must remember to give final answers to the appropriate number of significant figures, as discussed in chapter 2.

Suppose that, having completed the calculations, we find that we have overlooked something. Before taking the natural logarithm of the counts we should have subtracted the contribution to the counts due to the background radiation. This would require a fair amount of extra work if we were doing it by hand; however, the steps required are easily performed by our spreadsheet. We simply begin a new column and label it ln(Counts — background) as shown in screen 9.6. The background was 42 counts and so the formula entered into cell **D2** is LN(B2-42).

Screen 9.6: Entry of column allowing background subtraction before the taking of logs

	A	B	C	D
1	Thickness (mm)	Counts	ln(Counts)	ln(Counts — background)
2	1	1213	7.100851909	= LN(B2-42)
3	2	981	6.88857246	

As before, when the enter key is pressed, the calculation is carried out; the result is shown in screen 9.7.

Screen 9.7: Cell D2 set up to calculate ln(counts — background) automatically

	A	B	C	D
1	Thickness (mm)	Counts	ln(Counts)	ln(Counts — background)
2	1	1213	7.100851909	7.065613364
3	2	981	6.88857246	

Using the Fill Down command, the formula is copied from cell **D2** into all the cells down to cell D13, as shown in screen 9.8.

Screen 9.8: Fill down command used in the D column

	A	B	C	D
1	Thickness (mm)	Counts	ln(Counts)	ln(Counts — background)
2	1	1213	7.100851909	7.065613364
3	2	981	6.88857246	6.844815479
4	3	752	6.622736324	6.56526497
5	4	631	6.447305863	6.378426184
6	5	423	6.047372179	5.942799375
7	6	491	6.196444128	6.107022888

166

SCREEN 9.8 (continued)

	A	B	C	D
8	7	329	5.796057751	5.659482216
9	8	259	5.556828062	5.379897354
10	9	215	5.370638028	5.153291594
11	10	158	5.062595033	4.753590191
12	11	141	4.94875989	4.59511985
13	12	118	4.770684624	4.33073334

The greater the complexity and number of operations to be performed on a set of data, and the more data there are, the more a spreadsheet becomes the natural choice for data analysis.

We can see that though a spreadsheet does little that cannot be done 'by hand', laborious and repetitive calculations are done away with. This gives us more time to consider what the results really mean.

9.2.2 Built-in functions

As well as the natural logarithm function used in the previous section, spreadsheets have built into them all the mathematical functions that you would normally find on a scientific calculator such as SIN, COS, TAN and EXP. Statistical functions are available such as mean and standard deviation. To use these functions you must indicate the cells that contain the numbers to which the function is to be applied and specify the cells in which the calculated values should appear.

Screen 9.9 shows an example of four very useful functions generally available with spreadsheets. Seventy-seven data values representing the breaking strength (in newtons) of a batch of carbon fibres are shown. Beneath the columns of numbers are the values of the maximum and minimum breaking strengths, and the mean and standard deviation[7] of the values contained in the cells A2 through to G12. Each of these quantities has been calculated by specifying two things: the range of cells containing the numbers to be included in the calculation (rows 2 through to 12, and columns A through to G) and the particular statistical function to be applied.

---

7. This is the estimate of the population standard deviation, s, discussed in chapter 5.

Screen 9.9: Example of use of four of the built-in statistical functions

	A	B	C	D	E	F	G
1	Strength of carbon fibres (N)						
2	107	87	153	117	104	158	100
3	81	126	99	112	125	110	114
4	189	166	113	185	142	108	125
5	145	117	168	137	97	120	139
6	139	133	151	134	110	121	89
7	110	119	108	117	125	130	126
8	189	89	125	154	102	107	101
9	135	155	178	115	145	158	126
10	109	128	77	112	129	111	142
11	126	104	122	126	81	116	110
12	140	180	107	98	131	147	136
13							
14	max =	189		mean =	126		
15	min =	77		std.dev =	25		

Although a spreadsheet is normally laid out as rows and columns of equally sized small rectangles, the layout of the spreadsheet can be varied to make it easier to read. In screen 9.9 a large heading has been added to improve the clarity of the spreadsheet. In general, it is sensible to take advantage of these presentation options, as rows and columns filled only with numbers can be very difficult to comprehend later.

9.3 Graphing capabilities of a spreadsheet

No matter how important the data are that you have entered into a spreadsheet, their presentation as rows and columns of numbers is visually dull, and interesting features such as outliers or gaps in the data may be overlooked. It is better to have a pictorial representation of data and the professional spreadsheets on the market offer a huge variety of options for presenting data in graphical form. It is perhaps this facility of spreadsheets above all others, that is so attractive for our purposes. If, in your experiment, you have gathered, say, 100 pairs of x-y values, the thought of plotting them by hand on graph paper can be very discouraging. A spreadsheet can handle hundreds, if not thousands, of data points and plot them in under a second. With a click of a button the data and graph can be transferred to a printer so that a paper copy of the work can be obtained.

Figure 9.2 shows the data in screen 9.9 presented in the form of a histogram. The spreadsheet has ordered the data and counted the number occurring in each interval appearing along the x-axis.

Spreadsheets really come into their own when there are many data values. Figure 9.3 shows an X-ray diffraction pattern of a compound containing titanium and oxygen. In total there are 500 data values which were recorded by the X-ray equipment and then transferred directly into the spreadsheet so that they could be plotted and analysed.

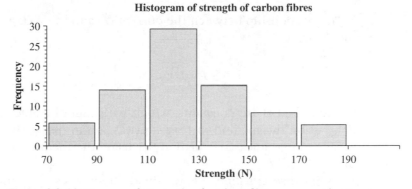

Figure 9.2: Histogram of strength of carbon fibres prepared using a spreadsheet

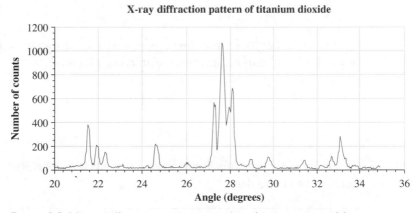

Figure 9.3: X-ray diffraction pattern produced using a spreadsheet

9.4 Example of the application of a spreadsheet

In this section we will indicate the power of a spreadsheet to analyse data by considering a specific example in some detail. Table 9.1 shows data that have been gathered in an experiment to investigate the electrical characteristics of a semiconductor device.[8] As the voltage applied to the device increases, so the current changes.

Table 9.1: Voltage and current values for a semiconductor device

Voltage (mV)	5	10	15	20	25	30	35	40	45	50
Current (mA)	124	147	133	108	82	60	42	29	20	13

The relationship between the current (I) and the voltage (V) for the device is:

$$I = DV \exp\left(\frac{-V}{B}\right) \tag{9.1}$$

where D and B are constants, which we should be able to determine using the method of least squares. Before using the spreadsheet, we first linearise equation 9.1 using the methods discussed in section 3.3.5.

To linearise equation 9.1, we divide both sides by V then take natural logarithms to give:

$$\ln\left(\frac{I}{V}\right) = -\frac{V}{B} + \ln(D) \tag{9.2}$$

This equation is of the form $y = mx + c$, where $y = \ln\left(\frac{I}{V}\right)$, $x = V$. The gradient, m, and intercept, c, are given by:

$$m = -\frac{1}{B} \text{ and} \tag{9.3a}$$

$$c = \ln(D). \tag{9.3b}$$

Screen 9.10 contains the values necessary to perform a least-squares fit to the data.

---

8. The device is called a tunnel diode.

Screen 9.10: Spreadsheet set up to perform a least-squares fit of a straight line to data

	A	B	C	D	E	F
1		$x = V$ (mV)	I (mA)	$y = \ln (I/V)$	xy (mV)	x^2 (mV)^2
2		5	124	3.210844	16.05422	25
3		10	147	2.687847	26.87847	100
4		15	133	2.182299	32.73448	225
5		20	108	1.686399	33.72798	400
6		25	82	1.187843	29.69609	625
7		30	60	0.693147	20.79442	900
8		35	42	0.182322	6.381254	1225
9		40	29	−0.32158	−12.8633	1600
10		45	20	−0.81093	−36.4919	2025
11		50	13	−1.34707	−67.3537	2500
12	Sum =	275	758	9.351114	49.55802	9625
13						

Row 12 shows the sums of each column of numbers. We see that:

$$\sum x_i = 275 \text{ mV}, \quad \sum y_i = 9.351114, \quad \sum x_i y_i = 49.55802 \text{ mV and}$$

$$\sum x_i^2 = 9625 \text{ (mV)}^2$$

These numbers are required for linear regression. The formulae for the gradient, m, and the intercept, c, using unweighted linear regression were given in chapter 6. They are:

$$m = \frac{n \sum x_i y_i - \sum x_i \sum y_i}{n \sum x_i^2 - \left(\sum x_i\right)^2}$$

$$c = \frac{\sum x_i^2 \sum y_i - \sum x_i \sum x_i y_i}{n \sum x_i^2 - \left(\sum x_i\right)^2}$$

where n is the number of data pairs.

We can enter the previous equations directly into cells **B13** and **B14** in the spreadsheet as shown in screen 9.11.

Screen 9.11: Formulae entered into cell B13 and B14 to calculate gradient and intercept respectively

	A	B	C	D	E	F	G
12	Sum =	275	758	9.351114	49.55802	9625	29.64071
13	m =	−0.1007					
14	c =	3.7031					

Using equations 9.3a and 9.3b we can calculate B and D:

$$B = -\frac{1}{m} = -\frac{1}{-0.1007} = 9.930 \text{ mV, and}$$

$$D = \exp(c) = \exp(3.7031) = 40.572 \text{ A V}^{-1}$$

Figure 9.4 shows a plot of the data, after the linearisation has occurred. The graph was drawn using the plotting facilities of the spreadsheet.

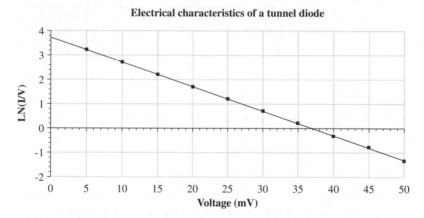

Figure 9.4: Graph showing ln (I/V) as a function of voltage

Any modification to a data point will cause the gradient and intercept to be automatically recalculated. In addition, the line of best fit on the graph will be recalculated and displayed if any data are changed.

If a *weighted* least-squares fit to the data were required, then an extra column would need to be inserted into the spreadsheet containing the uncertainty in each observed y-value. Table 6.6 in section 6.3.1 shows a possible layout of a spreadsheet used to perform a fit using weighted least squares.

9.5 Other features offered by spreadsheets

Although this chapter has introduced spreadsheets mainly with regard to their use in analysing and presenting x-y data, the scope of their features extends far beyond this. Options offered by sophisticated packages (such as Excel) include the manipulation of complex numbers, data smoothing and Fourier analysis. Here we will describe some other features common to spreadsheets which are useful for the analysis of experimental data.

9.5.1 'What if' calculations

A 'what if' calculation is no more than an application of the techniques we have already discussed concerning the modification of data in spreadsheets. Suppose we require the answers to the following questions.

In a least-squares problem, what happens to the gradient, m, and intercept, c, (and the uncertainties in these quantities) if the uncertainties in the individual y-data values:
 (i) increase?
 (ii) decrease?
(iii) are proportional to the magnitude of the y-quantity?

Additionally, what happens to m and c if an outlier is added or removed?

These represent questions which, using a pocket calculator, could take hours of work to answer. However, once a spreadsheet is set up to perform a weighted least-squares fit to data then data values can be changed or removed and the uncertainties likewise modified, and the consequences seen immediately.

9.5.2 Transferring data into a spreadsheet

Another important ability of spreadsheets is to take in data that has already been gathered or 'logged' by a data acquisition system. Some experiments generate so much data — for example like that shown in the X-ray diffraction pattern in figure 9.3 — that it is not possible to record all the data 'by hand'. In such circumstances, the data are logged by a piece of equipment (often controlled by computer). The data are stored in a file on a disk and then transferred to a spreadsheet for analysis and presentation. A common way of storing data on a disc is in a 'text' file in which data are stored in the form of ASCII characters.[9] A spreadsheet such as

---

9. ASCII stands for American Standard Code for Information Interchange. This code is widely used for data transference.

Excel can transfer the contents of a text file into its cells (this is referred to as 'importing' a file) for analysis. As characters may be encountered which are additional to those representing the experimental data, a good spreadsheet has editing facilities to remove those characters.

9.5.3 Sorting numbers

The sorting of numbers (in ascending or descending order) is a standard feature on all spreadsheets. Before a histogram can be plotted, for example, data are sorted so that the number of data values falling into predetermined intervals can be quickly counted.

9.5.4 Linear least squares with more than two unknown parameters

If we wish to extend our linear regression analysis to include situations where there are more than two unknown parameters, then we can use the spreadsheet's facility for solving matrices. So long as a function is linear in the unknown parameters — for example:

$$y = ax^2 + b \ln (x) + c$$

where a, b and c are the unknown parameters, it is possible to perform a linear least squares fit.[10] The problem is that as the number of parameters increases, so does the algebraic complexity of the solution for those parameters. However, using the matrices facility of a spreadsheet allows us to solve for the unknown parameters in a matter of seconds.

9.5.5 Non-linear least squares

One of the most powerful facilities that spreadsheets can offer is the ability to solve problems in situations where an analytical solution is not possible. When an equation is not linear in the unknown parameters such as:

$$y = a \exp (bx) + c \qquad\qquad (9.4)$$

linearisation is not possible and so linear least squares cannot be used to solve for a, b and c. However, using a spreadsheet, we can solve for a, b and c iteratively. The technique involves setting up the sum of squares, $(y_{io} - y_{ic})^2$, in a cell on the spreadsheet and

---

10. For a discussion of linear regression extended to more than two parameters, see *A First Course in Linear Regression* by M. S. Younger (details in appendix 1).

then instructing the spreadsheet to minimise that sum by varying the adjustable parameters (which for equation 9.4 would be the parameters a, b, and c). The program proceeds to vary the parameters until the sum of squares is minimised and then prints out their final values. In principle, this facility allows us to extend our least squares fitting to take account of any function.[11]

9.6 Alternatives to spreadsheets

Spreadsheets are attractive due to their ease of use, extensive 'on-line' help[12] and their availability. As they have been designed to accommodate a wide range of users such as those from business and commerce, not all the features that scientists and engineers desire are yet found in a spreadsheet. There are packages designed specifically for the analysis and presentation of scientific and technical data which offer extra options to those found on spreadsheets, such as numerical differentiation and integration of plotted data, factor analysis, cross-tabulation tables and non-parametric tests. If you need extra data analysis features then there are many commercially available packages on the market, one of which is StatView.[13]

The gap between the best spreadsheet packages and those designed specifically for the analysis of scientific and technical data is narrowing and it may be that, for all but the most specialised requirements, spreadsheets will become the preferred option.

9.7 Comment

Pencil, paper and a pocket calculator are often adequate when you wish to analyse a small amount of data. However, as microcomputers and good spreadsheet packages become more widely available, they will be favoured when large amounts of data are involved. For many people the spreadsheet is already the first choice for routine data analysis tasks. The increasing miniaturisation of computers means that powerful spreadsheets are available on 'palm-top' computers making them almost as portable as a laboratory notebook and a pocket calculator.

---

11. This type of fitting is also referred to as non-linear regression. Refer to *A First Course in Linear Regression* (see appendix 1) for details.
12. 'On-line' help refers to instructions available on screen of the microcomputer which, in many cases, can be as extensive as the user manual supplied with the spreadsheet package.
13. StatView is a package for the Macintosh computer and is available from Abacus Concepts, Berkeley, California.

Some care is required when using a spreadsheet. Like a laboratory notebook, to use a spreadsheet most effectively requires a certain amount of organisation. Spreadsheet files can easily be mislaid if not carefully catalogued. Data stored magnetically on a disk can become corrupted (or even subject to computer viruses) so care must be taken to ensure that copies are made of important files.

Despite these minor drawbacks, the power, flexibility and ease of use of a spreadsheet makes it an excellent tool for data analysis and spreadsheets are destined to play an increasingly important role in the analysis and presentation of data in both research and teaching laboratories.

APPENDIX 1:
CHAPTER REFERENCES AND FURTHER READING

This appendix contains details of books referenced in each chapter. Other texts which go into more detail of the topics discussed in this book are also recommended for general reading.

Chapter references

Chapter 2

Rocke, F. A. (1984), *Handbook of Units and Quantities*, Australian Atomic Energy Commission, N.S.W.

Includes a comprehensive description of the SI system of units, prefixes and conversion factors between systems of units. There is a particularly useful section which contains worked examples using SI units.

Chapter 4

Kaye, G. W. C. and Laby, T. H. (1986), *Tables of Physical and Chemical Constants* (15th edition), Longman, New York.

An excellent reference book for physical, chemical and astronomical data. It includes an introduction to units and their definitions as well as a short introduction to the statistical treatment of data.

Chapter 5

Meyer, S. L. (1975), *Data Analysis for Scientists and Engineers*, John Wiley & Sons, New York

A large and comprehensive text, most appropriate to study at a level beyond first year college or university. Deals with advanced topics in data analysis.

Taylor, J. R. (1982), *An Introduction to Error Analysis,* Oxford University Press, California

A well-written book dealing in detail with a wide range of matters concerning uncertainties. It includes topics not discussed in this book, such as the binomial distribution, covariance, and hypothesis testing.

Chapter 6

Bevington, P. R. and Robinson, D. K. (1992), *Data Reduction and Error Analysis for the Physical Sciences* (2nd edition), McGraw-Hill, New York

Established advanced text which includes computer routines (in PASCAL) to assist data analysis. Contains a good section dealing with non-linear least-squares fitting of functions to data.

Taylor, J. R. (1982) *An Introduction to Error Analysis,* Oxford University Press, California

(See comments under chapter 5.)

Younger, M. S. (1985), *A First Course in Linear Regression* (2nd edition) Duxbury Press, Boston

Discusses the technique of least squares in some depth (though not designed specifically with scientists and engineers in mind). The book includes details of a number of computer packages with least-squares analysis features.

Chapter 7

Lindsay, D. (1984), *A Guide to Scientific Writing,* Longman, Melbourne

A small book containing very sound advice on writing laboratory reports, long reports and articles for publication. It has a 'biological sciences' flavour, but much of the advice and suggestions on scientific writing are applicable to all students of science and engineering.

Chapter 9

Dodge, M., Kinata, C. and Stinson, C. (1993), *Running Microsoft Excel for Windows,* (4th edition), Microsoft Press, Washington

A large and detailed text on the use of the Excel spreadsheet package. Written in a 'user friendly' manner with many examples, and begins assuming no prior knowledge of Excel. However, some idea of what a spreadsheet *is* is assumed.

Younger, M. S. (1985), *A First Course in Linear Regression* (2nd Edition) Duxbury Press, Boston

(See comments under chapter 6.)

General further reading
Experimentation and apparatus

Holman, J. P. (1984), *Experimental Methods for Engineers* (4th edition) McGraw-Hill, New York

A wide-ranging text dealing mainly with experimental techniques and instruments of measurement, such as pressure, flow, temperature and force measurements. Many diagrams, graphs and tables describing the characteristics of particular transducers are included.

Moore, J. H., Davis, C. C. and Coplan, M. A. (1989), *Building Scientific Apparatus* (2nd edition) Addison-Wesley, California

Much more than a 'do-it-yourself' guide to building apparatus. Presents basic principles in areas such as vacuum systems, properties of materials, optical technology and practical electronics. Contains a large number of useful graphs (for example summarising optical data). Also included are details of suppliers and distributors of equipment (though details are limited to contact points in the United States).

Data analysis and presentation

Coleman, H. W. and Steele, W. G. (1989), *Experimentation and Uncertainty Analysis for Engineers,* John Wiley & Sons, New York

A senior undergraduate text concentrating on experimental design and particularly on the detailed analysis of experimental uncertainties. Many practical examples of uncertainty assessment are discussed in detail. Though SI units appear in the book, liberal use is made of other systems of units.

Pugh, E. M. and Winslow, G. H. (1966), *The Analysis of Physical Measurements,* Addison-Wesley, Massachusetts

Advanced (up to final year undergraduate) treatment of the statistical analysis of data. Many useful problems (with answers!) drawn from various areas of the physical sciences are included.

APPENDIX 2:
DEGREES OF FREEDOM AND THE t DISTRIBUTION

Degrees of freedom

It was stated in chapter 5 that the estimate of the population standard deviation, s, is given by:

$$s = \left(\frac{\sum (x_i - \bar{x})^2}{n-1} \right)^{\frac{1}{2}}$$ **(5.7)**

The denominator of the equation ($n - 1$) is equal to the number of *degrees of freedom* in the calculation (the symbol usually used to represent this quantity is v). The reason that v is one less than the number of data is that we have placed a small restriction on the values that x_i can take. This arises from the fact that in order to calculate the population standard deviation, we first need to know the population mean, and the best we can do is to use the mean of the sample of data, $\bar{x}$, as an estimate of the population mean. So, for example, if the mean of five numbers is 7.3 and four of the numbers are, 8.7, 5.3, 6.2 and 7.9, the last data value *must* be 8.4. In this situation, one degree of freedom has been 'lost' because although (in this example) four of the five numbers can have any value, the fifth cannot.

A similar situation arises when calculating an estimate of the population standard deviation, σ, of x-y data points about a straight line through the points. In chapter 6 this equation was given by:

$$\sigma = \left[\frac{1}{n-2} \sum (y_i - mx_i - c)^2 \right]^{\frac{1}{2}}$$ **(6.14)**

Here the number of degrees of freedom is $n - 2$ due to the fact that *two* restrictions are placed on the possible values of y because two quantities are calculated from these values, namely the gradient, m, and the intercept, c.

In general, whenever a population parameter, such as the population standard deviation, is estimated using sample data drawn from that population, one degree of freedom is lost for every quantity appearing in the formula, which itself is calculated using the sample data.

The t distribution

When we calculate the confidence limits for the true value of a quantity based on a sample drawn from a population (as we do when we make repeat measurements of a particular quantity) the width of normal distribution, given by the standard deviation, is an underestimation of the spread in the data values. In fact, the smaller the sample (i.e. the smaller the number of repeat measurements) the larger is the spread. This has an influence on our calculations for confidence limits, though as we round confidence limits to one significant figure, the influence is usually small.

The statistical distribution we use to describe the spread of small samples of data is called the t distribution (or, sometimes, Student's t distribution). It is very similar to the normal distribution in that it is a symmetrical distribution which 'tails away' as the deviation of a value from the mean increases. Figure A2.1 shows the normal distribution and the t distribution.

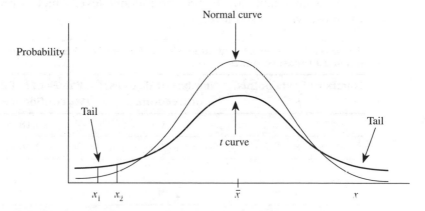

Figure A2.1: Normal and t probability distributions

The t curve is flatter than the normal curve and there is more area in the tails beneath the curve. For both curves the area beneath the curve between any two values of x (for example x_1

and x_2 in figure A2.1) is proportional to the probability that a value of x will be observed within that interval. We can see that the area between x_1 and x_2 is greater for the t curve compared to the normal curve. This means that, according to the t distribution, there is a greater probability that a value x will be observed in this interval (which is far from the mean) in comparison to the normal distribution.

Whereas the shape of the normal curve is independent of the number of data points, n, the t curve becomes flatter as the number of points decreases. For large values of n, the t curve and the normal curve are indistinguishable.

The confidence limits for the true value of a quantity in situations where the number of repeat measurements is small can be written:

$$\text{confidence limits for true value of a quantity} = \bar{x} \pm t_c \sigma_{\bar{x}} \qquad \textbf{(A2.1)}$$

where $\sigma_{\bar{x}} = \dfrac{s}{\sqrt{n}}$

The value s is given by equation 5.7 and t_c is a number which depends on the number of data points and the degree of confidence required. Textbooks concentrating on statistics of data, such as that by Meyer (see appendix 1), contain tables of t_c for a wide range of confidence levels and number of data points. Table A2.1 shows a small table for the 95% confidence level which is probably the most widely used level.

Table A2.1: Values of t_c at the 95% confidence level for between 4 and 30 measurements

Number of data points, n	Number of degrees of freedom, v	Values of t_c for the 95% confidence limits
4	3	3.18
8	7	2.36
12	11	2.20
20	19	2.09
30	29	2.05

As an example, suppose we want the 95% confidence limits for the true value of the following repeat measurements of the mass of a body (units kg) 6.5, 8.0, 7.4, 7.2.

Using these data we have:

$$\bar{x} = 7.275 \text{ kg}$$
$$s = 0.6185 \text{ kg}$$
$$\sigma_{\bar{x}} = 0.3093 \text{ kg}$$

Using equation A2.1 and by referring to table A2.1 we see that for $n = 4$ the 95% confidence limit value of t_c is 3.18, so that:

$$\text{true value} = (7.275 \pm 3.18 \times 0.3093) \text{ kg}$$

$$= (7 \pm 1) \text{ kg}$$

If we had used the normal distribution, we would have found the 95% confidence limit to be (7.3 ± 0.6) kg. When quoting confidence limits to one significant figure, the difference between using the normal and t distribution is usually negligible when n exceeds ten.

APPENDIX 3:
COMBINATION OF INDEPENDENT UNCERTAINTIES

Suppose a quantity V depends on a and b so that we can write $V = V(a,b)$. If the uncertainty in the measurement of quantity a in no way depends on the uncertainty in the measurement of quantity b, we can determine the uncertainty in V using the ideas about quantifying the variability in data discussed in chapter 5.

We begin by writing the variance of V as (see equation 5.1):

$$\sigma_V^2 = \frac{\sum (V_i - \bar{V})^2}{n} = \frac{\sum (\Delta V_i)^2}{n} \qquad \textbf{(A3.1)}$$

As V depends upon a and b, we can write:

$$\Delta V_i = \frac{\partial V}{\partial a} \Delta a_i + \frac{\partial V}{\partial b} \Delta b_i$$

Substituting this into equation A3.1 gives:

$$\sigma_V^2 = \frac{\sum \left(\frac{\partial V}{\partial a} \Delta a_i + \frac{\partial V}{\partial b} \Delta b_i \right)^2}{n} \qquad \textbf{(A3.2)}$$

The right-hand side of equation A3.2 can be expanded to give:

$$\sigma_V^2 = \sum \frac{\left(\frac{\partial V}{\partial a} \Delta a_i \right)^2}{n} + \sum \frac{\left(\frac{\partial V}{\partial b} \Delta b_i \right)^2}{n} + \sum \frac{2 \frac{\partial V}{\partial a} \frac{\partial V}{\partial b} \Delta a_i \Delta b_i}{n} \qquad \textbf{(A3.3)}$$

The first two terms on the right-hand side of the equation will always be greater than zero and so cannot be neglected. By contrast, the product $\Delta a_i \Delta b_i$ in the third term will sometimes be positive and other times negative, depending on the sign of Δa_i and Δb_i. To a good approximation, the last term will be zero with positive terms in the summation cancelling out the negative terms. We are left with:

$$\sigma_V^2 = \left(\frac{\partial V}{\partial a}\right)^2 \sum \frac{(\Delta a_i)^2}{n} + \left(\frac{\partial V}{\partial b}\right)^2 \sum \frac{(\Delta b_i)^2}{n}$$

We can write:

$$\sigma_a^2 = \sum \frac{(\Delta a_i)^2}{n} \text{ and } \sigma_b^2 = \sum \frac{(\Delta b_i)^2}{n}$$

so that:

$$\sigma_V^2 = \left(\frac{\partial V}{\partial a}\right)^2 \sigma_a^2 + \left(\frac{\partial V}{\partial b}\right)^2 \sigma_b^2 \qquad \text{(A3.4)}$$

We have shown that, for measurements in which the uncertainties are independent, the variance in the calculated value is equal to the sum of the variances in the measured quantities.

It is the standard error of the mean of a set of repeated measurements that we take to be the uncertainty in a quantity. In order to calculate the standard error in $\bar{V}$, $\sigma_{\bar{V}}$, when the standard error in $\bar{a}$ and $\bar{b}$ are given by, $\sigma_{\bar{a}}$ and $\sigma_{\bar{b}}$ respectively, σ_V in equation A3.4 is replaced by $\sigma_{\bar{V}}$, σ_a by σ_a and so on. Equation A3.4 now becomes:

$$\sigma_{\bar{V}}^2 = \left(\frac{\partial V}{\partial a}\right)^2 \sigma_{\bar{a}}^2 + \left(\frac{\partial V}{\partial b}\right)^2 \sigma_{\bar{b}}^2 \qquad \text{(A3.5)}$$

Therefore the uncertainty in V, $\sigma_{\bar{V}}$, can be written:

$$\sigma_{\bar{V}} = \sqrt{\left(\frac{\partial V}{\partial a}\right)^2 \sigma_{\bar{a}}^2 + \left(\frac{\partial V}{\partial b}\right)^2 \sigma_{\bar{b}}^2}$$

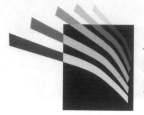

APPENDIX 4:
DATA GATHERING USING A MICROCOMPUTER

A4.1 Overview: how can a microcomputer make measurements?

We saw in chapter 9 that, with appropriate software, a micro-computer can assist in the efficient analysis of experimental data. The versatility of the microcomputer is such that, as well as being a powerful tool for data analysis, it can play a central role in systems that can *gather* data.

The benefits of using a microcomputer to gather data include:

(i) the ability to record changes in physical quantities which span very short to very long time intervals (say between fractions of a second and months)

(ii) immediate display of data in tabular or graphical forms

(iii) storage of large amounts of data in a form readable by data-analysis packages such as a spreadsheet

(iv) data gathering without requiring the continuous attention of the experimenter

(v) the facility to take *control* over certain aspects of an experiment, such as turning on power supplies and starting and stopping motors.

A4.1.1 Microcomputers handle numbers

The first thing to recognise is that computers deal internally with numbers. Whatever physical quantities are to be measured during an experiment they must ultimately be represented within a computer as numbers. In an experiment we might want to measure quantities as diverse as temperature, force, electrical current or pH. How can such quantities be 'translated' into numbers?

There are usually three steps involved: the first is to produce a voltage (or current) which is proportional to the physical quantity being measured using a *transducer*. The second stage is to modify the voltage if it is too large or too small so that it is ready to pass on to the third stage. Voltages produced by transducers are often termed *signals* and the modification process referred to as *signal conditioning*. The third stage is to use an electronic device referred to as an *analogue to digital converter*[1] or ADC for short. The purpose of the ADC is to convert a voltage to a number. After a number has been generated, it can be stored into the memory of the computer.

A typical system for gathering data is shown in the form of a block diagram in figure A4.1.

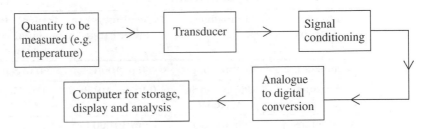

Figure A4.1: Block diagram of a typical computer-controlled data-gathering system

In this appendix we will consider transducers, signal conditioning, ADCs and other matters relevant to data gathering using a micro-computer.

A4.2 Transducers

In order to monitor a physical quantity with a computer we first need a means of generating a voltage (or current) which is proportional to the quantity being studied. A phrase commonly used by scientists and engineers to describe this process is to 'convert' a pressure, or light intensity and so on, to a voltage.

Table A4.1 gives a list of commonly used physical transducers along with the typical change in the signal voltage or current you can expect for a given change in the physical quantity.

Looking at the last column in table A4.1 we see that some transducers generate voltages in the millivolts range and others in the volt range. Similarly, some transducers generate currents spanning

---

1. The term 'analogue' voltage refers to the fact that, within certain limits, say 0 to 5 V, a voltage can take on any value.

μA to mA. How can a measurement system cope with such large ranges of voltage and current? In a conventional experiment in which a multimeter is used to measure voltages and currents, the problem is easily dealt with. By simply turning a knob or pressing a button on the meter we are able to select a range which allows us to measure the voltage or current with the desired resolution. Unless we are lucky enough to have a multimeter that can be controlled directly by the microcomputer[2], we need to find a way to match the output voltage or current of the transducer to the stage where a voltage is converted to a number.

Table A4.1: Commonly used transducers

Physical quantity	Transducer	Typical measurement range	Typical change in signal voltage or current
Temperature	Thermocouple	0 to 100°C	0 to 5 mV
Pressure	Piezoelectric material	0 to 3×10^5 Pa	0 to 250 mV
Light intensity	Photodiode	10 to 10 000 Lx	0.1 μA to 1 mA
Magnetic field	Hall effect probe	0 to 40 mT	2 to 6 V
Position or displacement	Precision potentiometer	0 to 200 mm	0 to 10 mV

A4.3 Signal conditioning: a little electronics goes a long way

ADCs to be discussed in the next section permit the best resolution of a voltage when the output of the signal-conditioning section of the data-gathering system matches the input range of the ADC. A typical input range for the ADC is 0 to 5 V. Suppose that, during an experiment, a transducer generates a voltage in the range 0 to 70 mV. How can we convert 0 to 70 mV to 0 to 5 V?

The simple answer is that we need a voltage amplifier which has an 'amplification factor', more usually termed *voltage gain*, of:

$$\frac{5 \text{ V}}{70 \text{ mV}} = 71 \text{ (to two significant figures)}$$

---

2. Multimeters that can 'communicate' with a computer are fairly expensive. It is usual to find instruments with such a facility costing in excess of $1000.

Though amplifiers can be bought 'off the shelf' for the above purpose, they tend to be expensive. A cost-effective way of signal conditioning is to use a small electronic device called an *operational amplifier.*

A4.3.1 The operational amplifier

The operational amplifier, or 'op-amp' for short, is a multipurpose electronic device which may be adapted for many situations requiring signal conditioning. Most op-amps come in a small rectangular plastic package with eight terminals, generally referred to as 'pins'. Figure A4.2 shows the layout of the pins for the AD548 op-amp (manufactured by Analog Devices). Pins 1 and 5 are used to provide small adjustments to the output of the op-amp,[3] therefore we will not deal with their use here.

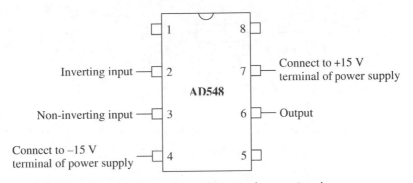

Figure A4.2: Layout of connections of a typical operational amplifier

All op-amps have two inputs and an output, and most will operate from a power supply which provides voltages which are positive and negative with respect to a designated 'zero volts' terminal.

A very important attribute of any signal-conditioning system is that, once the conditioning is attached to the transducer, the output from the transducer should not be affected. So, for example, if we attach an amplifier to a transducer and this causes the voltage from the transducer to decrease from 70 mV to 60 mV, then the amplifier is having an adverse effect on the output of the transducer and the circuit requires modification.

For the output voltage of a transducer to remain unaffected when it is connected to signal-conditioning circuitry, the input

---

3. Pin 8 has no electrical function at all.

resistance of the signal-conditioning circuitry must be at least a few orders of magnitude larger than the output resistance of the transducer. The AD548 op-amp shown in figure A4.2 has an input resistance of about 10^{13} Ω, which is large enough for all but the most demanding of applications.

A4.3.2 Example of use of an op-amp

We will show briefly here how an op-amp such as the AD548 shown in figure A4.2 can be used to provide a gain of about 70 to a signal from a transducer.

The AD548, in common with most other op-amps, requires a power supply capable of supplying +15 V and −15 V (dc) with respect to its zero volts terminal.[4]

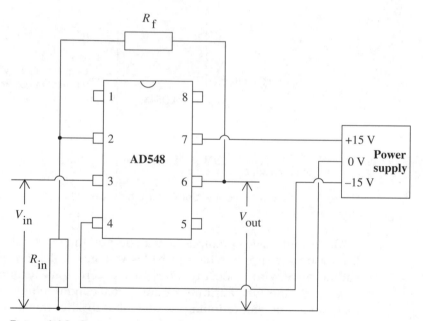

Figure A4.3: Op-amp connected as an amplifier

Pins 1, 5 and 8 do not play a part in this circuit and are left without any connection. V_{in} refers to the voltage coming from the transducer, and V_{out} is the voltage after signal conditioning has occurred. Two resistors, R_f and R_{in} have been placed in the circuit

---

4. The current required to operate an op-amp is usually of the order of tens of milliamps, and most power supplies are capable of providing this.

and their function is to set the gain of the amplifier. The relationship between V_{in} and V_{out} is:

$$V_{out} = \left(1 + \frac{R_f}{R_{in}}\right)V_{in} \qquad \textbf{(A4.1)}$$

To complete the design, if we choose R_f to be 83 kΩ and R_{in} to be 1.2 kΩ then the voltage gain which is defined as:

$$\text{voltage gain} = \frac{V_{out}}{V_{in}}$$

would be equal to (by rearranging equation A4.1):

$$\text{voltage gain} = \left(1 + \frac{R_f}{R_{in}}\right) = 1 + \frac{83 \text{ k}\Omega}{1.2 \text{ k}\Omega}$$

$$\approx 70 \text{ (gain has no units)}$$

The gain of about 70 means that a voltage produced by a transducer of, say, 30 mV, will be increased to about 2.1 V.

A4.4 The analogue to digital converter

Before a microcomputer can store or analyse voltages from a transducer, an analogue to digital converter must be used which accepts any voltage within a specified range on its input side (usually 0 to 5 V or 0 to 10 V), and on its output side generates a positive whole number proportional to the input voltage. Because, at their most fundamental level, all microcomputers represent numbers in binary, we hear terms such as 8-bit ADCs, 12-bit ADCs and 16-bit ADCs. The number of 'bits' refers to the number of binary digits used to represent the voltage converted by the ADC. An 8-bit ADC, for example, can produce a binary number from 00000000 to 11111111 which, when converted to decimal becomes,[5] 0 to 255.

So, for example, if we were to use an 8-bit ADC which had an input range of 0 to 5 V, then the ADC would produce the number '0' for 0 V and the number 255 for 5 V. The smallest increase in voltage that the ADC could detect would be:

$$\frac{5 \text{ V}}{255} = 19.6 \text{ mV}$$

We can say that the 8-bit ADC has a resolution of 19.6 mV per bit. A 12-bit ADC is able to output any number from 0 to 4095

---

5. If required, some pocket calculators (such as those in the CASIO range) will convert numbers from binary (base 2) to decimal.

(in decimal) so that if it had the same input range as the 8-bit ADC discussed above, its resolution would be:

$$\frac{5\ V}{4095} = 1.22\ mV\ per\ bit$$

The greater the number of bits of the ADC, the greater is its resolution. However, there is a price to pay for higher resolution and that is the time taken for the ADC to complete the conversion of the voltage to a number. The greater the number of bits, the longer the time for conversion to be completed. The most common types of ADC available have 12-bit resolution and can usually complete a conversion in under 50 μs. At a cost there are ADCs with more bits and shorter conversion times. However, many of the data-gathering situations encountered in science and engineering do not require the extra resolution and speed.

ADCs are not usually found as standard items in a micro-computer but are available on an electronic circuit board which can usually be placed within the case of the microcomputer. When such a board has built on it other items which are useful when performing an experiment (such as an electronic timer), it is some-times referred to as a *multipurpose interface card*.

EXERCISE

Assuming you had a 16-bit ADC which could accept voltages in the range 0 to 10 V, what would be the resolution of this ADC?

A4.4.1 Multipurpose interface cards

In the context of computers, *interfacing* refers to the steps that must be taken so that a computer can acquire data, control equipment or communicate with other electronic devices. Manufacturers of microcomputers rarely include with their machine the necessary electronics, usually referred to as *hardware*, to permit interfacing. However, there are many companies which supply hardware to allow the interfacing of microcomputers to experiments.

In addition to one or more ADCs, a typical multipurpose inter-face card (MIC) includes a number of other useful circuits such as digital to analogue converters and digital input and output.[6]

---

6. There are many multipurpose interfacing cards available for the two most popular types of microcomputer, the Macintosh and the IBM compatibles. One popular card for the IBM compatibles, is the advanced interfacing board supplied by Sunset Laboratory, 19th Avenue Forest Grove, Oregon, USA.

As the name suggests, a digital to analogue converter or DAC for short, converts a number (presented to it in binary form) into an analogue voltage. One or more DACs are used in situations where the microcomputer is required to exercise fine control over a circuit or device. An example of the use of a DAC would be in the control of an *x-y* plotter used to plot graphs. By connecting one DAC to the *x* input and another to the *y* input of the plotter, a microcomputer can control the position of the pen of the plotter precisely.

ADCs and DACs are not required in situations where the input from a device or output to a device can only take on two values. For example, if an experiment requires the timing of an object moving between two light detectors, we would not need to know the exact value of the light intensity being detected at any instant. A digital input to a microcomputer can be used to register a '0' when a light detector is fully illuminated and a '1' when an object passes momentarily over the detector.

In an experiment which requires a vacuum pump, electrical motor, or lamp to be switched on, a microcomputer can provide the 'on-off' control via digital output lines. The digital output has two states which correspond to 0 V and 5 V (or voltages close to these values). Such voltages themselves may not be able to switch equipment on and off, but one way this can be achieved is by connecting the digital output to an electromechanical relay which in turn does the switching.

Table A4.2 summarises the types of circuits found on an MIC along with an example of their application to experimentation.

Table A4.2: Types of circuits found on a multipurpose interface card

Type of circuit	Typical use in experimentation
Analogue to digital converter	Convert voltages from transducers which detect smoothly varying quantities (e.g. temperature) to digital values
Digital to analogue converter	Provides fine control over voltages or currents supplied to external circuits (e.g. control of electrical current to a heater)
Digital input and output	input: for counting events (e.g. counting number of X-rays received by an X-ray detector) output: switching equipment on and off (e.g. opening and closing a valve to control the flow of a liquid)
Timer	Timing events (e.g. the time between heartbeats)

A4.5 Controlling the data-gathering process

Automated data gathering requires that a microcomputer carry out all the tasks that would normally be done manually. Amongst other things, the microcomputer needs to be instructed when to make measurements and how many to make in total. It must be responsible for manipulating the data (for example converting ADC values into something more meaningful, such as values in SI units). A microcomputer needs step-by-step instructions on what to do and when to do it. The instructions are more commonly referred to as a computer *program* and for data-gathering purposes a useful program might take anywhere from a few minutes to months to write.

As a preliminary to writing a computer program, it is necessary to have clear in our own minds what exactly we expect the computer to do. It can be helpful to write out in detail the steps that would be taken manually when performing an experiment.

As an example, in an experiment involving the measurement of temperature every 10 s over a period of 30 minutes, the following steps would be taken.

(i) Start stopwatch, measure and record temperature.

(ii) Wait for 10 s to elapse.

(iii) Measure and record temperature.

(iv) Go to step (ii) unless 30 minutes in total have elapsed, in which case stop the experiment.

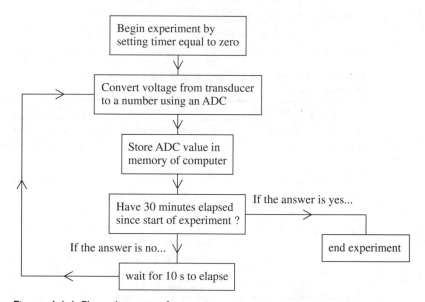

Figure A4.4: Flow diagram of a temperature-measuring experiment

To translate this into a computer program it can sometimes be useful to begin by drawing up a diagram which includes all the necessary steps.[7] Such a diagram is referred to as a *flow* diagram. Figure A4.4 shows a flow diagram for the temperature measurement discussed above.

There are basically three main options when going from the flow diagram shown in figure A4.4 to a working set of instructions that the microcomputer can understand. We will refer to them as assembly language, high level programming language and dedicated interface software.

A4.5.1 Assembly language

When a computer performs any task, such as adding or storing numbers, fetching numbers generated by an ADC or responding to the pressing of buttons on a keyboard, it does so because it has instructions stored in its memory. Instructions are given in a language constructed from 'ones' and 'noughts', that is, a binary language. In terms of physical quantities the 'ones' and 'noughts' are voltages (usually 5 V and 0 V respectively). Ultimately *any* set of instructions, no matter how they are generated, must be turned into binary before they are executed. Because it is difficult for humans to work in binary (try multiplying 1011000100011111 by 1101101101111101) we need methods by which both simple and complex instructions can be turned into binary on our behalf.

Assembly language is made up of simple instructions, which when combined can perform very complex tasks. However, the small section of a typical assembly language program shown below is enough to reveal one difficulty with this approach to program writing:

```
mov cx,ffff
mov al,01
out dx,al
mov al,00
out dx,al
dec cx
```

Let us look at the first line of this section of program and find out what it is doing. Within the microprocessor there are special memory locations set aside which are used over and over again to

7. We assume that a temperature transducer has been chosen and that signal conditioning has occurred.

perform simple tasks. Sixteen consecutive locations are termed a *register* and in the IBM compatibles, one of those registers is given the name 'cx'.

The ffff in the first line of the program is a number expressed not in binary or decimal, but *hexadecimal*. Hexadecimal numbers (known as 'hex' for short) are numbers in the base 16 and are frequently used because of the convenience of translating from binary to hex.[8] For example when all '1's fill a register, this number in hex is ffff.

Finally, looking at the beginning of the first line of the program, we see the word, 'mov'. mov instructs the computer to take the number ffff and place it in the cx register.

A fair amount of explanation has been required for a fairly simple task. Abbreviated explanations can be placed alongside each instruction and these can help to explain what the program is doing. Once the program has been completed, it is translated or 'assembled' into a form understood by the microcomputer. It is at this stage that 'bugs' in the program are identified and the program is returned to the programmer for attention.

From this brief introduction, it might seem that to write programs in assembly language requires much hard work. However, there are two significant rewards to be had from this approach with regard to data gathering from experiments:

(i)   Programs written in assembly language run very quickly. If data gathering from an experiment has to be very rapid — say, for example, you are trying to record a transient effect that only lasts milliseconds — you might have to 'capture' the data using an assembly language program.

(ii) Assembly language programs are very efficient in their use of memory space inside a microcomputer. This might be important if you wish to store a large amount of data from an experiment and need to use every available memory location in the microcomputer.

One important point to note is that as an assembly language program is tied to the type of microprocessor being used in the computer, it cannot easily be transferred to another type of machine. For example, an assembly program developed for an IBM compatible computer will have to be rewritten for an Apple Macintosh computer.

---

8. The relationship between decimal and hexadecimal is:

Decimal	0	1	2	3	4	5	6	7	8	9	10	11	12	13	14	15	16
Hex	0	1	2	3	4	5	6	7	8	9	a	b	c	d	e	f	10

A4.5.2 High-level languages

Long assembly language programs are sometimes difficult to write and debug. When short, fast routines are required, they might be the first programming choice. However, if a program is required to gather data from an experiment, manipulate some numbers and store the data onto a disc, assembly language is normally replaced by a high-level language, such as BASIC, Pascal or C. When using a high-level language for general programming purposes, talk of 'binary' and 'registers' disappears as knowledge of these matters is less important than when close to the machine level. However, when dealing with interfacing, knowledge of the more fundamental elements of a computer is still important.

All high-level languages allow you to do the same things as in assembly language, but permit programs to be structured in a way that makes them easier to read, change and debug. The BASIC language[9] has been adopted by many experimenters involved with interfacing, so we will restrict our discussion to this language.

The BASIC programming language includes many instructions which are recognisable as English words such as PRINT, FOR, TO, NEXT and END. These words are used to form a BASIC program, for example:

```
10 FOR I = 1 TO 10
20 PRINT I, I*I
30 NEXT I
40 END
```

First we note that the program has numbers to the left of each of the instructions. These are termed 'line numbers'. The computer normally carries out the instructions by working down the list a line at a time but the line numbers allow the programmer to jump out of sequence if required.

The letter 'I' in the line numbered 10 is used as a counter and the program will first set I equal to one. Line 20 tells the computer to print out (onto the screen of the monitor attached to the computer) the number 1 and 1 multiplied by 1. Line 30 acts to increase I by one and send the computer back to line 20 to print 2 and 2 multiplied by 2. The process continues until I = 10 whereupon the instruction at line 40 would be executed, which would stop the program.

---

9. BASIC stands for Beginners All purpose Symbolic Instruction Code.

In this example, line 20 instructs the computer to simply print out numbers. However, it could equally instruct the computer to start a conversion on an ADC connected to a temperature transducer, and then return the ADC value to the program and print this out on the screen. If we then included a BASIC instruction which required the computer to wait ten seconds before it gathered another ADC value, we would have many of the elements shown in the flow diagram in figure A4.4 for temperature measurement in a real experiment.

BASIC and other computer languages have error-trapping routines which assist greatly when you are trying to write a program. Nevertheless, data-gathering programs generally require that you become fairly familiar with the programming language. There is an approach to data gathering using a microcomputer that takes some of the work out of learning what happens at the machine level and does not require the experimenter to learn a programming language.

A4.5.3 Dedicated interface software

The growth in the use of microcomputers as tools for gathering data in experiments has seen the expansion in the number of interface cards offering both general and specialist data acquisition options. Alongside the growth in number of interface cards available has been software of varying levels of 'user friendliness' which is designed specifically to exploit the data capture and control options of the cards. It is possible, for example, with the right combination of card and software, to turn a microcomputer into a sophisticated oscilloscope.[10] The combination of hardware and software has the capability to capture rapidly varying voltage waveforms, 'freeze' and display them. All the options that you would normally find on an oscilloscope, such as variable time base, addition and subtraction of waveforms appearing on two channels and variable voltage gain, are built into the software.

Some of the most sophisticated software for data gathering allows you to create an instrument dedicated to your experiment by choosing images of voltmeters, ammeters, switches and so on and arranging them on the screen of the microcomputer. Once the program has been told of the logical connection between the images on the screen and the interface cards in the computer and the appropriate connections made to the transducers, everything

---

10. Software 'snapshot' and hardware CYDAS16 multipurpose interface card supplied by Boston Technology, Union Street, North Sydney NSW 2060 has such a capability.

can be run and be monitored from the screen, which in effect acts as the 'front panel' of the instrument. An arrangement like this is sometimes referred to as a 'virtual instrument'.

Many dedicated interface software packages have built-in data analysis tools, such as least-squares fitting and graph plotting. Those that do not have these facilities allow for gathered data to be stored on a disk in a form that can be read into an analysis package, such as a spreadsheet.

A4.6 Other interface options

More and more laboratory equipment, such as voltmeters, spectrometers and digital balances are supplied complete with an interface to allow easy communication with a microcomputer. This does away with the need for a multipurpose interface card, because, essentially, the instrument makes the measurement (as instructed by the microcomputer) and passes along the value from that measurement to the microcomputer. The two most popular types of interfaces available to control external instruments are the GPIB (standing for the General Purpose Interface Bus) and the RS232. As for the multipurpose interface cards, some 'ready-made' software is available to run instruments through the GPIB and RS232 interfaces. It is possible to design an experiment in which instruments measuring some physical quantities are controlled through a GPIB interface while at the same time a multipurpose interface card takes care of timing or other experiment-related functions.

A4.7 Comment

In this appendix we have touched upon the application of microcomputers to data gathering. Due to their flexibility and cheapness, it is likely that microcomputers will play an ever-increasing role in the gathering as well as the analysis and presentation of data derived from experiments. Full appreciation of the power and limitations of microcomputers used in this context requires that we be familiar with the performance and characteristics of transducers, signal-conditioning circuits, interface cards and interface software. Textbooks on the subject of data gathering using microcomputers, and instruction manuals provided by manufacturers of interface cards are of great help when getting started. The suggestions for further reading below may be of assistance to those who wish to devise their own computer-based data-gathering system.

A4.8 Further reading

In the areas of microcomputers and electronics, developments happen very rapidly and some sections of a text can be out of date before a book is published. However, the small selection of texts mentioned in this section include much information that should be of lasting value.

Cripps, M. (1989), *Computer Interfacing: Connection to the Real World,* Edward Arnold, London

Good general book dealing with the issues of transducers, signal conditioning and interfacing. Also deals with connection of a microcomputer to instruments through standard interfaces.

Horowitz, P. and Hill, W. (1989), *The Art of Electronics* (2nd edition), Cambridge University Press, Cambridge

Considered by many to be *the* book on all aspects of analogue and digital electronics. It is extremely well written and contains many examples of applications of electronics.

Sheingold, D. H. (1981), *Transducer Interfacing Handbook,* Analog Devices, Massachusetts

Though this book deals exclusively with signal-conditioning devices manufactured by the company publishing the book, there is a wealth of practical advice that only those who are deeply immersed in the business of practical electronics can furnish. General principles are followed up with practical circuits.

ANSWERS

CHAPTER 2

Exercises

A

(i) 2.2 μV,   (ii) 62 mm,   (iii) 65.2 kJ,   (iv) 0.18 MW,   (v) 67 pF

B

Pressure ($\times 10^5$ Pa)	1.03	1.01	1.01	0.99	1.05	1.08

C

(i) three,   (ii) two,   (iii) three,   (iv) four,   (v) three

D

(i) 18.9, (ii) 0.108, (iii) 725, (iv) 1.76, (v) 62 700

E

(i) 10,   (ii) 562,   (iii) 0.000 24,   (iv) 469,   (v) 0.185,   (vi) 71.4

F

1. (i) 5.654×10^{-3},   (ii) 1.250×10^2,   (iii) 9.384×10^7,
   (iv) 3.400×10^6,   (v) 1.001×10^{-7}
2. (i) 5.7×10^{-3},   (ii) 1.3×10^2,   (iii) 9.4×10^7,   (iv) 3.4×10^6,
   (v) 1.0×10^{-7}

Problems

1. 46.9 g
2. 52×10^{-3} kg
3. The student has inserted the diameter of the sphere into formula for the volume instead of the radius, omitted the units and has given the volume and the density to too many

significant figures. Inserting the radius gives the volume as 56.5 mm³. The value for the density of the sphere must be changed, given to the correct number of significant figures (two, if we adopt the rule given in section 2.5.2) and the units must be inserted. The density of the sphere turns out to be 7.8×10^{-3} g mm⁻³.

CHAPTER 3

Exercises

A

Faults:

(i) The graph shows time *versus* temperature, not temperature versus time.

(ii) The units have been omitted from the x-axis.

(iii) The data point gathered at temperature 42°C has been omitted.

(iv) The last data point has been incorrectly plotted.

B

1. (i) 0.0044 T   (ii) 0.0090 T

2.

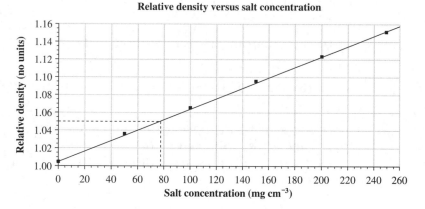

From the line on the graph, when the relative density is 1.053, the corresponding salt concentration is about 78 mg cm⁻³.

C

My 'best line' through the points gave a gradient, m, of 0.026 Ω°C⁻¹. I found the coordinates of one point on the line to be $x = 75$°C and $y = 7.15$ Ω Using equation 3.6, the intercept

(where the temperature is 0°C), $c = 7.15 - 0.026 \times 75 = 5.2\ \Omega$
The assumption made is that the linear change of resistance with temperature continues down to 0°C (that is, that extrapolation is valid).

D

What to plot	gradient, m, =	intercept, c, =
F versus N	μ	0
v versus t	a	u
$\dfrac{R}{T}$ versus T	B	A
$\ln(I)$ versus x	$-\mu$	$\ln(I_0)$
T versus $\sqrt{l}$	$\dfrac{2\pi}{\sqrt{g}}$	0
$\dfrac{1}{v}$ versus $\dfrac{1}{u}$	-1	$\dfrac{1}{f}$
H versus T	C	$-CT_0$
$\ln\left(\dfrac{I}{V}\right)$ versus V^2	$-B$	$\ln(A)$
T_w versus R^2	$-k$	T_c

E
$\log_{10}(A) = -10.4$, so $A = 3.98 \times 10^{-11}$, $n = 3.96$

Problems

Note: The answers given here are based on lines that I have drawn 'by hand' through x-y data points. When comparing your answers with those given here, remember that it is unlikely that your lines will coincide exactly with mine so that slight variations can be expected in values calculated for gradient and intercept.

1. (ii) Gradient = 6.09 N kg⁻¹ and intercept = 0.02 N. Therefore, we can write $y = 6.09x + 0.02$
    (iii) (a) 4.28 N
          (b) 7.94 N

(iv) I found the maximum gradient ≈ 6.7 N kg^{-1} and minimum gradient ≈ 5.6 N kg^{-1}, so that the gradient can be written as, $m = (6.1 \pm 0.6)$ N kg^{-1}. The maximum intercept ≈ 0.44 N and the minimum ≈ -0.48 N, so that the intercept can be written as, $c = (0.0 \pm 0.5)$ N.

2. (i) $m = 322.6$ Pa K^{-1}
   (ii) c (found by choosing a point on the best line, then using equation 3.6) $= 1.930 \times 10^3$ Pa
   (iii) At $T = 330$ K, $P = 1.08 \times 10^5$ Pa. At $T = 400$ K, $P = 1.31 \times 10^5$ Pa.
   (iv) $T = 304$ K
3. (iii) mercury concentration $= (1.30 \pm 0.15)$ ppb
4. (i) To obtain straight line, plot $\dfrac{1}{I}$ versus R

   (ii) gradient, $m = \dfrac{1}{E}$ and intercept, $c = \dfrac{r}{E}$
5. (i) Take the natural logarithms of both sides of the equation then plot ln (R) versus ln (I).
   (iii) $m = 1.49, c = 3.15$
   (iv) $n = 1.49, k = 23.3$
6. (i)

Concentration (M)	Reaction rate (s^{-1})
0.20	0.0400
0.16	0.0333
0.14	0.0313
0.12	0.0250
0.08	0.0179
0.02	0.0086

(iii) Gradient of line $= 0.179$ M^{-1} s^{-1}

CHAPTER 4

Exercises

A

1. In an experiment I performed using an electric kettle which shuts off automatically when the water boils, I recorded the following times it took for 600 cm^3 of water to begin to boil.

Time (s)	135	115	119

The first measurement was made with the whole kettle 'cold'. After the first boiling, the average temperature of the kettle increased and so when further water was added some heat will have been transferred from kettle to water, even before the kettle was switched on. This would have reduced the amount of heat required from the element to heat the water to boiling point, thus reducing the time to reach this point. Other factors that could have affected the heating time were:

(i) Variability in the thermostat, that is, it does not always switch the kettle off at the same temperature every time.

(ii) The temperature of the tap water could have changed between experiments.

(iii) The amount of water added each time could have been slightly different.

2. In an experiment in which a coin was dropped from a distance of 3 m, I recorded the following data.

Time (s)	0.80	0.66	0.80	0.83	0.78	0.82	0.82	0.80	0.84	0.87

Largest value 0.87 s, smallest value 0.66 s.

Coinciding the starting of the watch with the fall of the coin, and reacting to the coin hitting the ground seem to be the most important factors affecting the timing. This could easily introduce an uncertainty of about 0.2 s to any individual measurement.

B

1. 46 cm^3, $3.83 \times 10^{-7} \text{ m}^3 \text{ s}^{-1}$
2. $4.2 \times 10^{-3} \text{ Pa}$

C

1. 0.018 s
2. Mean = 425 mm and uncertainty = 4 mm

D

(i) Too many figures in mean *and* uncertainty — should be $(12.6 \pm 0.6) \text{ N m}^{-1}$

(ii) Too many figures in mean, and would be better to write this in scientific notation, that is $(8.8 \pm 0.5) \times 10^3 \text{ kg m}^{-3}$

(iii) No units given! — they should be m s^{-1}

(iv) Mean should be quoted to same number of decimal places as the uncertainty, that is $(1.100 \pm 0.001) \text{ μF}$

(v) ± 1 does not convey anything meaningful. It *could* mean $\pm 1 \times 10^{-7} \text{ m}$, but it is not clear.

E

(i) Fractional uncertainty = 0.1, percentage uncertainty = 10%
(ii) Fractional uncertainty = 0.04, percentage uncertainty = 4%
(iii) Fractional uncertainty = 0.3, percentage uncertainty = 30%
(iv) Fractional uncertainty = 0.11, percentage uncertainty = 11%

F

1. $a = 6.0$ m s^{-2}, $a_{max} = 6.6$ m s^{-2}, $a_{min} = 5.4$ m s^{-2}, $\Delta a = 0.6$ m s^{-2}.
2. $v = 46$ m s^{-1}, $v_{max} = 50$ m s^{-1}, $v_{min} = 42$ m s^{-1}

G

(i) $\dfrac{\partial s}{\partial a} = \tfrac{1}{2}t^2$ and $\dfrac{\partial s}{\partial t} = at$

(ii) $\dfrac{\partial P}{\partial I} = 2IR$ and $\dfrac{\partial P}{\partial R} = I^2$

(iii) $\dfrac{\partial n}{\partial i} = \dfrac{\cos i}{\sin r}$ and $\dfrac{\partial n}{\partial r} = \dfrac{-\sin i \cos r}{\sin^2 r}$

(iv) $\dfrac{\partial v}{\partial T} = \tfrac{1}{2}\left(\dfrac{1}{\mu T}\right)^{\frac{1}{2}}$ and $\dfrac{\partial v}{\partial \mu} = -\tfrac{1}{2}\left(\dfrac{T}{\mu^3}\right)^{\frac{1}{2}}$

H

1. $B = (1.61 \pm 0.12) \times 10^{-3}$ T   3. $f = (12.0 \pm 0.2)$ cm
2. $j = (1.04 \pm 0.12) \times 10^6$ A m^{-2} 4. $N = (7.7 \pm 1.3) \times 10^3$

Problems

1. (i) Range = $1 \times 10^{-2}\,\Omega$
    (ii) Mean = $9.44 \times 10^{-2}\,\Omega$, uncertainty in resistance
        $= 1.3 \times 10^{-3}\,\Omega$
    (iii) resistance of mercury sample = $(9.44 \pm 0.13) \times 10^{-2}\,\Omega$
2. $n = (0.49 \pm 0.13)$ mol
3. (i) Flow rate = (0.437 ± 0.013) m s^{-1}
        pH = 7.04 ± 0.07
        Temperature = $(10.5 \pm 0.3)°C$
        Electrical conductivity = $(9.4 \pm 0.5) \times 10^2$ µS cm^{-1}
        Lead content = (52 ± 5) ppb
    (ii) Largest fractional uncertainty is in the lead content
        (fractional uncertainty = 0.1)
4. (i) $k = 6.9$ kg s^{-2} (note that the spring constant is normally
        expressed in units N m^{-1} which is equivalent to kg s^{-2})
    (ii) $k = (6.9 \pm 1.4)$ kg s^{-2}
    (iii) $k = (6.9 \pm 1.4)$ kg s^{-2}

5. Using the partial differentiation method to determine the uncertainty (see section 4.5.3.1), $G = (11.4 \pm 0.7)$ dB

6. (i) $\dfrac{\partial z}{\partial a} = e^b$, $\dfrac{\partial z}{\partial b} = ae^b$ (ii) $z = (1.24 \pm 0.18) \times 10^5$

7.

uncertainty in $\left(\dfrac{1}{\text{reaction time}}\right)$ (s^{-1})
0.003
0.002
0.002
0.0019
0.0010
0.0002

CHAPTER 5

Exercises

A

1. $\bar{x} = 24.8125$, $\sigma = 1.7011$, $\sigma_{\bar{x}} = 0.6014$, so we would write the best estimate of the true value as (24.8 ± 0.6).
2. $\bar{x} = 175.9$ mm, variance $= 5.010$ mm^2, $\sigma = 2.238$ mm, $\sigma_{\bar{x}} = 0.2889$ mm. We can write the best estimate of the rebound height as (175.9 ± 0.3) mm.

B

1.

Interval (mm)	Frequency
$145 \leq x < 150$	2
$150 \leq x < 155$	2
$155 \leq x < 160$	11
$160 \leq x < 165$	13
$165 \leq x < 170$	16
$170 \leq x < 175$	9
$175 \leq x < 180$	6
$180 \leq x < 185$	1

207

3. 38 values lie between $\pm\sigma$ of the mean
57 values lie between $\pm 2\sigma$ of the mean
   The normal distribution predicts about 70% of values between $\pm\sigma$ of the mean and 95% between $\pm 2\sigma$ of the mean. 70% of 60 is 42 and 95% of 60 is 57.

C

(i) $\bar{x} = 75.67, \sigma = 8.219, s = 8.585$
(ii) $\sigma_{\bar{x}} = 2.373$ (using σ) $\sigma_{\bar{x}} = 2.478$ (using s)
(iii) $\bar{x} = 76 \pm 2$

D

1. $5 \times 10^{-4}\,\text{g mm}^{-3}$
2. (i) $n = 1.421$

   (ii) $\dfrac{\partial n}{\partial i} = \dfrac{\cos i}{\sin r} = 0.7875, \dfrac{\partial n}{\partial r} = \dfrac{-\sin i \cos r}{\sin^2 r} = -1.818$

   (iii) $\sigma_{\bar{n}} = \sqrt{\left(\dfrac{\partial n}{\partial i}\right)^2 \sigma_i^2 + \left(\dfrac{\partial n}{\partial r}\right)^2 \sigma_r^2}$

   (iv) Note $1° \equiv 1.745 \times 10^{-2}$ rad
   $\sigma_{\bar{n}} = 0.06$

   (v) $n = 1.42 \pm 0.06$ (refractive index has no units)

Problems

1. (i) $\sigma = 0.26, s = 0.37$
   (ii) using first three numbers, $\sigma = 0.22, s = 0.26$
   using first four numbers, $\sigma = 0.19, s = 0.22$ (same value to one significant figure)
   using first five numbers, $\sigma = 0.17$ and $s = 0.20$
2. (i) $\bar{x} = 9.11$ mg
   $\sigma^2 = 0.23\,(\text{mg})^2$
   $\sigma = 0.48$ mg
   (ii) $\sigma_{\bar{x}} = 0.14$ mg
   (iii) 70% confidence limits 8.97 mg to 9.25 mg
3. (ii) Difficult to tell — more data required
   (iii) mean time = 1.10 s, estimate of population standard deviation = 0.21 s, standard error of mean = 0.03 s. So time = (1.10 ± 0.03) s
   (iv) velocity = (320 ± 10) m s^{-1}
4. 95% confidence limits for the true number of particles = (151 ± 6)
5. Using method for combining uncertainties described in section 5.5: o.d. = (0.8 ± 0.2)

6. Using method for combining uncertainties described in section 5.5: $V_{in} = (6.3 \pm 0.3)$ mV
7. $\eta_f = (1.6 \pm 0.3)$ kg m^{-1} s^{-1}

CHAPTER 6

Exercises

A

1. (i) $m = 0.050$
   (ii) $m = 0.023$
2. $m = 7.315, c = 30.70$
3. (i) $m = -2.873 \times 10^{-6}$ s^{-2}, $c = 9.794$ m s^{-2}
   (ii) Multiply out the brackets to give $g' = g - \dfrac{2gd}{R_E}$. Comparing this with $y = mx + c$, we see that $c = g$, and $m = \dfrac{-2g}{R_E}$.

   From this it follows that, $g = 9.794$ m s^{-2}, and

   $R_E = \dfrac{-2g}{m} = 6.818 \times 10^6$ m. Incidentally, this compares with the 'accepted' value for the average radius of the Earth of $\sim 6.36 \times 10^6$ m.

B

(i) For question 2 of exercise A: $\sigma_m = 0.223$, $\sigma_c = 2.33$, so $m = (7.3 \pm 0.2)$ and $c = (31 \pm 2)$
(ii) For question 3 of exercise A: $\sigma_m = 9.2 \times 10^{-8}$ s^{-2}, $\sigma_c = 5.7 \times 10^{-3}$ m s^{-2} so, $m = (-2.87 \pm 0.09) \times 10^{-6}$ s^{-2} and $c = (9.794 \pm 0.006)$ m s^{-2}.

D

1. (iii) $m = -0.100\ 49$ s^{-1}, $c = 2.889$
   (iv) $\sigma_m = 0.0015$ s^{-1}, (so that $m = (-0.100\ 49 \pm 0.0015)$ s^{-1})
        $\sigma_c = 0.014$, (so that $c = (2.889 \pm 0.014)$)
   (v) $\tau = (9.95 \pm 0.15)$ s, $V_0 = (18.0 \pm 0.3)$ V
2. (ii) $I_0 = (2.88 \pm 0.09) \times 10^3$ counts s^{-1}, $\lambda = (1.00 \pm 0.02) \times 10^{-2}$ s^{-1}
3. (ii) $A = (0.290 \pm 0.013)$ mm^2, $B = (2.39 \pm 0.18)$ mm^2

Problems

1. $m = -1.1893, c = 77.508$
   $\sigma_m = 0.067, \sigma_c = 1.9$
   $m = -1.19 \pm 0.07$
   $c = 77.5 \pm 1.9$

ANSWERS

2. (i) Plot I versus $\cos 2\theta$. Gradient will be equal to $I_{max} - I_{min}$ and the intercept will be equal to I_{min}
    (ii) $m = 0.867$ and $c = 0.938$
    (iii) $\sigma_m = 0.03$ and $\sigma_c = 0.02$, so $m = (0.87 \pm 0.03)$ and $c = (0.94 \pm 0.02)$
    (iv) $I_{min} = 0.94$ and $I_{max} = 1.81$
3. (i) Squaring both sides gives $M^2 = kt + D$ therefore plotting M^2 against t should give a straight line of gradient k and intercept D.
    (ii) The uncertainty in M^2 is $2M\Delta M$, (see section 4.5.3.1), so the uncertainty in M^2 is not constant and a weighted fit is required.
    (iii) To three significant figures, $m = 61.4$ (mg)^2 h^{-1} and $c = 13.9$ (mg)^2.
4. Rearrange equation to give:
$$\frac{1}{g} = \frac{1}{np} + \frac{s}{n}$$
This is of the form $y = mx + c$ where $m = \frac{1}{n}$, and $c = \frac{s}{n}$
$c = 2204$ m^2 kg^{-1} $s = 1.589$ m^2 N^{-1}
$m = 1387$ N kg^{-1} $n = 7.210 \times 10^{-4}$ kg N^{-1}
5. (i) Rearranging the equation gives, $\frac{V-1}{\theta} = D\theta + B$.
    Therefore plotting $\frac{V-1}{\theta}$ against θ should give a straight line of gradient D and intercept B.
    (ii) $D = (-1.9 \pm 0.2)$ 10^{-6} cm^3 °C^{-2}
    $B = (3.219 \pm 0.013) \times 10^{-3}$ cm^3 °C^{-1}
6. (i) $I = I_0 \exp\left(\frac{eV}{nkT}\right)$, taking natural logs of both sides gives:
    $\ln(I) = \frac{eV}{nkT} + \ln(I_0)$, therefore plotting $\ln(I)$ against V should give a straight line of gradient $\frac{e}{nkT}$ and intercept $\ln(I_0)$.
    (ii) If $y = \ln(I)$, then $\Delta y = \frac{\partial y}{\partial I}\Delta I = \frac{1}{I}\Delta I$
    As $\frac{\Delta I}{I} = 0.02$ which is constant, an unweighted fit should be used.
    (iii) $m = 25.42$, $c = -22.12$, so that $I_0 = 2.47 \times 10^{-10}$ A and $n = 1.52$

7. (i) $t^2 = \dfrac{d^2}{v^2} + \dfrac{4b^2}{v^2}$. Plotting t^2 against d^2 should give a gradient

of $\dfrac{1}{v^2}$ and intercept of $\dfrac{4b^2}{v^2}$.

(ii) $m = 0.3237 \times 10^{-6}\,\text{s}^2\,\text{m}^{-2}, c = 1179 \times 10^{-6}\,\text{s}^2$
$\sigma_m = 0.0019 \times 10^{-6}\,\text{s}^2\,\text{m}^{-2}, \sigma_c = 2.5 \times 10^{-6}\,\text{s}^2$

(iii) $v = 1757\,\text{m s}^{-1}, b = 30.2\,\text{m}$

CHAPTER 7

Problem

Abstract

- Omit first line — this could be said in the introduction.
- Could include values obtained for g in the experiment (with uncertainties).
- What is the value of g reported elsewhere? Could be included here for comparison.

Introduction

- Could omit second sentence.

Materials and methods

- Too few details given about method — for example, how was the ball released in the 'free-fall' experiment?
- Some mention of reaction time when measuring fall time of ball required — reaction time isn't mentioned until method B is discussed and, even then, no estimate of its value is given.

Results

- No units are given in the third column of table 1.
- If the tape measure can be read to ±1 mm, then why is the uncertainty in distance quoted as ±10 mm?
- More information required as to how the uncertainty in the period of the pendulum was calculated.
- Graphs have too few points; that is, ideally more measurements required.
- Linear regression could have been used to find the gradient, and uncertainty in the gradient, of the best straight line through the points in figures 3 and 4.

Discussion

- Mention of the reaction time being *a small fraction of the total time measured* (in the pendulum experiment) but no details of this appear in the report.

Conclusion

• No important omissions or excess material.

References

• No mention of the page number in the reference to the book by Young.

CHAPTER 8

Exercises

A

$\bar{x} = 59.25$

B

$\sigma = 3.127\ 873\ 679 \times 10^{-2}$ (full precision) $= 3 \times 10^{-2}$ (one significant figure)

$s = 3.343\ 837\ 616 \times 10^{-2}$ (full precision) $= 3 \times 10^{-2}$ (one significant figure)

C

$\Sigma x = 2899$, and $\Sigma x^2 = 1\ 440\ 835$

D

$B = 194.440\ 124\ 1\ \Omega$, $A = 0.093\ 915\ 734$ V (full precision)

$B = 194\ \Omega$, $A = 0.0939$ V (three significant figures)

E

1. (i) $B = 105.308\ 262\ 6°C\ cm^{-1}$
    (ii) $A = 13.961\ 692\ 95°C$
    (iii) $\Sigma x = 38.3$ cm, $\Sigma x^2 = 229.73$ cm^2 and $\Delta = 370.95$ cm^2
    (iv) $\sigma^2 = 266.218\ 753\ 7°C^2$
    (v) $\sigma_B = 2.396\ 109\ 322°C\ cm^{-1}$, $\sigma_A = 12.840\ 161\ 66°C$
2. $B = (105 \pm 2)°C\ cm^{-1}$, $A = (14 \pm 13)°C$

APPENDIX 4

Exercise

Resolution $= 0.153$ mV per bit

212

INDEX